Towards Energy Sustainability

Towards Energy Sustainability

Editors

C.A. Brebbia
Wessex Institute, UK

J.J. Sendra
Universidad de Sevilla, Spain

WIT*PRESS* Southampton, Boston

Editors:

C.A. Brebbia
Wessex Institute, UK

J.J. Sendra
Universidad de Sevilla, Spain

Published by

WIT Press
Ashurst Lodge, Ashurst, Southampton, SO40 7AA, UK
Tel: 44 (0) 238 029 3223; Fax: 44 (0) 238 029 2853
E-Mail: witpress@witpress.com
http://www.witpress.com

For USA, Canada and Mexico

Computational Mechanics International Inc
25 Bridge Street, Billerica, MA 01821, USA
Tel: 978 667 5841; Fax: 978 667 7582
E-Mail: infousa@witpress.com
http://www.witpress.com

British Library Cataloguing-in-Publication Data

A Catalogue record for this book is available
from the British Library

Library of Congress Catalog Card Number: 2018932133

ISBN: 978-1-78466-319-3
eISBN: 978-1-78466-320-9

The texts of the papers in this volume were set individually by the authors or under their supervision. Only minor corrections to the text may have been carried out by the publisher.

Printed in Great Britain by Lightning Source, UK.

The material contained herein is reprinted from special editions of the International Journal of Energy Production and Management, Volume 2, Number 4 and Volume 3, Number 1 published by WIT Press.

Preface

The world's economy is fuelled by energy. Depletion of resources and severe environmental effects resulting from the continuous use of fossil fuels has motivated an increasing amount of interest in renewable energy resources and the search for sustainable energy policies.

The changes required to progress from an economy mainly focused on hydrocarbons to one taking advantage of sustainable energy resources require considerable scientific research as well as the development of new engineering systems. Energy policies and management are of primary importance to achieve the development of sustainability and need to be consistent with recent advances in energy production and distribution.

In many cases, the challenges lie as much in the conversion from renewable energies (wind, solar, etc) to useful forms (electricity, heat, fuel) at an acceptable cost (including damage to the environment), as in the integration of those resources into an existing infrastructure.

The diverse topics covered by these papers involved collaboration between different disciplines in order to arrive at optimum solutions; including studies of materials, energy networks, new energy resources, storage solutions, waste to energy systems, smart grids and many others.

The Editors are especially grateful to the reviewers, as well as to the authors for their contributions.

The Editors

Contents

SMART DUAL THERMAL NETWORK

VÍCTOR F. SÁNCHEZ[1], AMAIA URIARTE[1], ENERITZ BARREIRO[1] & MATTEO PORTA[2]
[1]Sustainable Construction Unit, Tecnalia, Spain.
[2]D´Appolonia, Italy.

ABSTRACT

Conventional district heating (DH) systems enable demand aggregation at district level and can provide high centralized heat generation performance values. However, thermal Renewable Energy Sources (RES) deployment at building level still remains low, and exploitation suboptimal, as it is limited by the instantaneous thermal load and storage capacity availability of each building. Buildings play the role of consumers that request a variable amount of heat over time and the thermal network the role of unidirectional heat supplier, without any smart interaction. The FP7 project A2PBEER has developed an innovative Smart Dual Thermal Network concept based on RES and Combined Heat and Power (CHP) as generation technologies, that enables transforming existing suboptimal DH systems, into integrated thermal networks with optimized performance and building level RES system production exploitation. It is based on an innovative Smart Dual Building Thermal Substation concept, which allows a bidirectional heat exchange of the buildings with the thermal network, and to aggregate district level distributed production and storage capacity (Virtual District Plant). With this approach buildings become prosumers maximizing decentralized RES production exploitation, as any possible local heat production surplus on any building of the district, will be delivered to the network to be used by other buildings. Additionally, this thermal network allows the delivery of the energy necessary to meet the heating and cooling demand of the buildings through a single hot water distribution network. In this way, it is possible to upgrade conventional DH systems to district heating and cooling systems, without the construction of a district cooling plant and a dedicated cooling distribution network. Cooling is produced at building level through sorption technologies using locally deployed solar collectors and the thermal network as energy sources. Finally, the district typologies and climatic conditions that maximize the potential of this thermal network concept have been identified.

Keywords: District heating, district cooling, distributed generation, smart grids, thermal substations.

1 INTRODUCTION

Conventional district heating (DH) systems enable demand aggregation at district level and, under appropriate technical and economical boundary conditions, can provide high centralized heat generation performance values and minimize the required aggregated generation capacity. Additionally, these systems offer the possibility to integrate very efficient generation technologies such as Combined Heat and Power (CHP). However, thermal Renewable Energy Sources (RES) deployment at building level still remains low, and exploitation suboptimal, as it is limited by the instantaneous thermal load and storage capacity availability of each building.

District heating solutions are common in many northern European countries with cold climatic conditions. However, district cooling systems even in countries with very hot climatic conditions that set the ideal frame to take full advantage of them, are not a usual approach. Instead, cooling generation is solved at building level through conventional compression chillers and reversible heat pumps.

The interaction between buildings and the thermal network is established by a strict hierarchy where buildings play the role of consumers of heat delivered at some specific conditions (temperature and pressure), and the district heating the role of unidirectional heat supplier connected through heating substations designed to enable such heat interaction. Thermal networks are managed according to reactive control strategies focused on adapting the status

DOI: 10.2495/EQ-V2-N4-315-326

of the system (plant generators, storage systems, pums, etc) to the evolving district demand, so that the supply temperature and pressure setpoints are met over time.

On the other hand, district heating and cooling systems are very unusual and, from a technical point of view, based on the deployment of dedicated heating and cooling plants and of specific distribution networks for heating and for cooling (4 pipe distribution systems). According to traditional approaches, the transition from currently existing district heating systems to advanced heating and cooling systems requires the installation of a district cooling plant and the deployment of a dedicated cooling distribution network. However, this approach must face huge technical and economic obstacles, as it requires a complicated construction process, with a strong impact on the exploitation of the buildings of the district and inconveniencies for building users for prolonged periods.

The objective of this study is to develop the innovative Smart Dual Thermal Network concept (SDTN) and the Smart Dual Building Thermal Substation, which is the key component to enable the implementation of the proposed concept. This innovative thermal network is based on RES and CHP as generation technologies, and aims at transforming existing suboptimal public DH systems, into heating and cooling thermal networks with optimized performance and building level RES system production exploitation, avoiding the limitations currently existing in traditional DH system upgrading procedures.

The Smart Dual Building Thermal Substation (SDBTS) allows a bidirectional heat exchange of the buildings with the thermal network, and to aggregate district level distributed production and storage capacity (Virtual District Plant). With this approach buildings become prosumers maximizing decentralized RES production exploitation, as any possible local heat production surplus on any building of the district, will be delivered to the network to be used by other buildings.

2 SMART DUAL THERMAL NETWORK CONCEPT (SDTN)

2.1 Smart Network. Bidirectional heat exchange.

Generally speaking, public districts will be formed by buildings with different usage characteristics and different potential for integration of solar thermal collectors and CHP units. Some buildings will operate on a yearly basis (office buildings, etc) while others might remain without any activity for prolonged periods (educational buildings during summer, etc), and the same differences could exist on a daily basis, as some buildings will remain occupied during working hours while others will only be occupied during nights.

Additionally, the characteristics of the energy demands of the buildings might also be very different in relation to the distribution on a daily and yearly basis, and also regarding the relative weight corresponding to heating, Domestic Hot Water (DHW) production and cooling. On the other hand, due to the architectonic features of the buildings (availability of roof space to deploy solar collectors, shading effects, etc) the RES deployment potential of each building might be very different.

Bidirectional heat exchange through Smart Dual Thermal Substation deployment, will enable the aggregation of the distributed solar collector, CHP and storage systems creating a Virtual District Plant that will allow taking advantage of possible synergies between the usage and the heat generation profiles of the different buildings, reducing the required capacities, improving generation efficiency, and increasing the operation intensity of the CHP units of the Virtual District Plant.

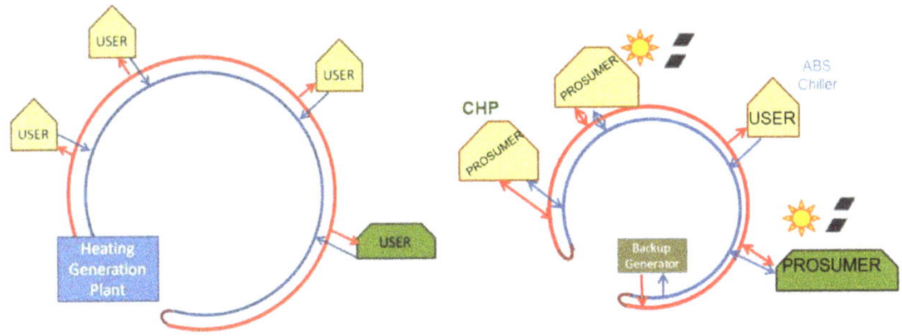

Figure 1: Conventional and innovative networks.

2.2 Dual Thermal Network

Additionally, this thermal network allows the delivery of the energy necessary to meet the heating and cooling demand of the buildings through a single hot water distribution network. In this way, it is possible to upgrade conventional DH systems to district heating and cooling systems, without the construction of a district cooling plant and a dedicated cooling distribution network. Cooling is produced at building level through conventional or innovative sorption technologies using locally deployed solar collectors and the thermal network as energy sources.

An innovative triple state sorption technology integrated into solar collectors has been developed by ClimateWell AB in the frame of the FP7 project A2PBEER, in order to provide this functionality. This collector can provide hot water (up to 95°C) operating in heating mode, and cool water (7°C) and low temperature hot water as a byproduct (35–45°C) operating in cooling mode [1]. In any case, conventional absorption chillers can also provide building level cooling production.

The nominal performance of conventional single effect absorption technologies is typically around 0.8, whereas for conventional compression based technologies higher performance values (2–6) can be obtained depending on the selected specific technology. Therefore, in order to make use of absorption as cooling generation technology in a competitive way in terms of efficiency and costs, the heat supply to absorption components is based on generation technologies that can provide free heat such as solar collectors, and technologies with very high economic performance such as CHP (when the cost of electricity is much higher than natural gas and other fossil fuels and biomass).

2.3 Integrated District Energy Management system based on model predictive control strategies.

In order to operate thermal networks as optimized integrated systems it is necessary to evolve from traditional reactive control approaches to a model predictive control approach [2]. Therefore, ICT management platform and energy model integration will be required.

The platform architecture must enable interoperability between building and district level energy management systems through standard communication protocols. This enables operating building level systems (CHP plants) according to district level constraints.

The developed framework is formed by a cluster of Building Supervisory Systems coupled to a District Supervisory System that will act as master. These systems are installed, respectively, on top of the existing Building Energy Management Systems (BEMS-s) and the district

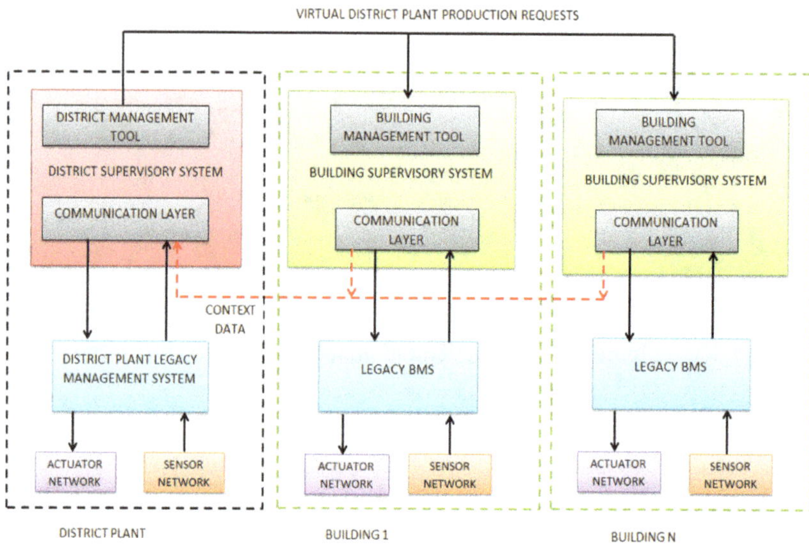

Figure 2: Architecture of the integrated district management system.

management system, and have the capacity to disseminate the required control commands to implement the optimized operational strategies at building and thermal network level.

2.4 Transition procedure

The thermal network concept has been developed for already existing public DH systems. Therefore, a flexible transition procedure that can be gradually implemented over time according to the technical constraints and economic availability of each specific project has been defined.

If the district heating plant is not based on CHP and RES, the process will consist in the gradual substitution of the existing heat generators, as they reach the end of their live span. In many cases, according to the existing specific constraints, the deployment of CHP will not take place at district plant level, but at Virtual District Plant level. The Virtual District Plant will be completed with the deployment of solar collectors at building level. Aggregation of distributed generation systems will increase the available flexibility to carry out the transition to the advanced thermal network concept. It is necessary to take into account that under many technical and economic circumstances, it might be easier to enable many small scale interventions distributed over several of the buildings connected to the network than big scale interventions at district heating plant level.

The transition procedure will be completed with the deployment of the Smart Dual Thermal Substation on the buildings of the district.

2.5 Technical and economic boundary conditions for district retrofitting through the SDTN

For the complete implementation of the SDTN concept the following technical and economic conditions must be fulfilled:

- Presence of relevant aggregated heating and cooling demands.
- The district should be formed by buildings with complementary energy use and local production patterns.

– Adequate balance between fuel and electricity prices.
– Favorable regulations regarding distributed production (feed in tariffs).
– Availability of an integrated district energy management system operating according to model predictive control strategies.

If not all these requirements are met, partial thermal network implementations will still be possible:

– The Smart Thermal Network (STN), including only the bidirectional heat exchange functionality (absence of cooling or unfavorable balance between fuel and electricity prices).
– The Dual Thermal Network (DTN) including only the dual nature of the network (absence of synergies between the demand and production profiles of the buildings of the district).

3 SMART DUAL THERMAL SUBSTATION

3.1 State of the art heating substations

A typical building heating substation has to fulfill several requirements, in order to ensure a satisfactory delivery of the required functionalities from the perspective of the consumer and the utility managers. An overview of such requirements is provided in Table 1.

In general, a building heating substation must provide energy to meet the needs of local heating systems and DHW production. The most usual substation types are the parallel systems and the 2 stage systems, in both cases with the possibility to incorporate a charging tank or instantaneous DHW production.

Parallel substations include specific heat exchangers for DHW production and for the heating system, connected in parallel to the thermal network. The 2 stage arrangement is more adequate for systems with a large DHW demand where the primary return flow of the heat exchanger of the heating system is used in the first stage of the DHW heat exchanger to preheat the makeup cold water. The temperature of the DHW is raised to the required level in the second stage of the DHW heat exchanger with hot water supply from the primary circuit of the thermal network. With this approach the return water temperature of the primary side of the DH is lowered further [3].

3.2 Smart Dual Thermal Substation functionalities, modules and typologies

The building thermal substation is the key component to transform existing traditional thermal networks into the developed advanced thermal network concept. Table 2 displays the required functionalities and modules.

Table 1: Conventional heating substation functional requirements.

General	Consumers	Utility managers
Accurate and stable temperature control	Sufficient supply temperature	Preferably indirect connection through heat exchangers
Low risk of bacteria growth	Low energy consumption	Low return temperature
Safe function	Accurate heat metering	Maximum flow limitation
Minimum space demand	Low noise level	Easy maintenance
Long operational life	Low cost	Sufficient differential pressure

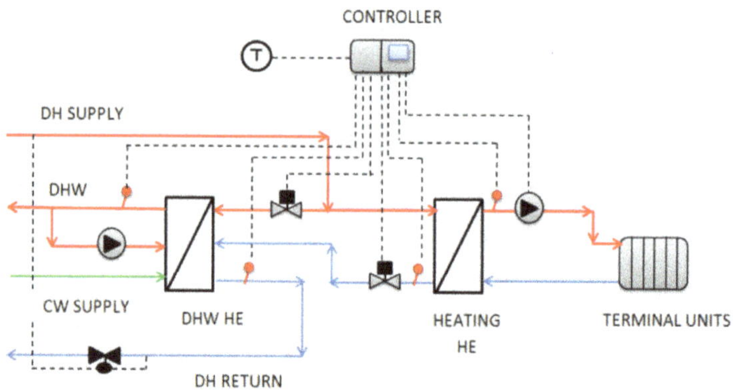

Figure 3: Two stage heating substation with instantaneous DHW production.

Table 2: Innovative thermal substation functionalities and modules.

Functionality	Description	Involved module
Heating	Heat delivery from the network for heating and DHW production	Heating module
Cooling	Heat delivery from the network for local cooling production through sorption components	Cooling module
Surplus delivery	Delivery of local production surplus	Surplus delivery module

According to the required functionalities and modules, the thermal substation typologies displayed in Table 3 have been defined.

Figure 4 displays the hydraulic scheme of the Dual Bidirectional Substation. As an evolution of the 2 stage arrangement used for state of the art heating substations [4], energy recovery from the outlet flow of the primary side of the cooling module is used as strategy to reduce the temperature value of the flow returned from the heating and cooling modules to the thermal network, minimizing distribution losses and improving the performance of the heat generators of the district plant.

The integration of the substation into the network is based on specific operational modes that provide the flexibility necessary to adjust the system to the requirements of all the foreseen interactions of the buildings and the network. Table 4 provides a summarized description of these operational modes, including the description of the operational settings.

Table 3: Innovative thermal substation typologies.

Typology	Functionalities	Required modules
Dual bidirectional	Heating, cooling and surplus delivery	Heating, cooling and surplus delivery
Dual unidirectional	Heating and cooling	Heating and cooling
Heating bidirectional	Heating and surplus delivery	Heating and surplus delivery

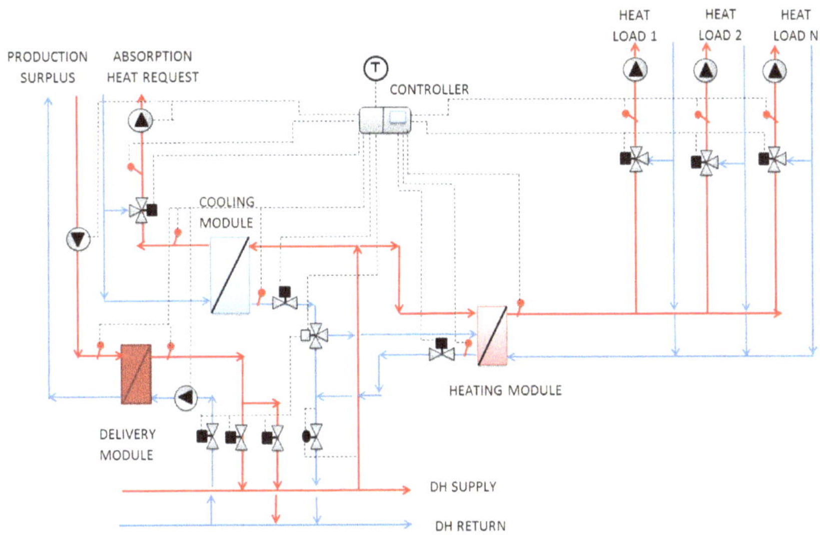

Figure 4: Smart Dual Thermal Substation hydraulic scheme

Table 4: Smart Dual Thermal Network operational modes.

Mode	Type	Functionality	Active modules	Season	Network setpoint (°C)	Surplus delivery
Heating	Network Local	Heating	Heating Delivery	Winter	65	Supply pipe (SP)
Cooling	Network Local	Cooling	Cooling Delivery	Summer	95	Return pipe (RP)
Heating and cooling	Network Local	Heating Cooling	Heating Cooling Delivery	Intermediate	95	RP
Surplus delivery	Local	Surplus delivery	Delivery	Permanently available	N.A.	SP RP

Regarding the heat delivery, two different energy delivery approaches have been defined to be activated according to the difference existing between the temperature value of the available local heat production surplus, and the temperature of the return and supply pipes of the thermal network.

If the temperature of the local heat production is higher than the supply temperature of the thermal network, the energy delivery will take place through the supply pipe (return/supply mode). However, if the temperature of the local hot water production surplus, is above the temperature of the return pipe of the network but below the network supply setpoint value, the energy delivery will take place through the return pipe (return/return delivery mode).

The return/return heat delivery mode will be less efficient, as the increase on the return temperature of the network will, in general, produce some decrease on the performance of the heat generators of the district plant, but the energy delivered to the thermal network will compensate this effect, and produce a net decrease on its fuel consumption.

4 THERMAL NETWORK ASSESSMENT METHODOLOGY

The evaluation of the potential of the SDTN is based on district dynamic behavior modelling and simulation. This is a complex goal as it involves systems affected by dynamic effects with very different time scales, contained inside space boundaries of very different scales, including technologies that are part of different domains.

Due to modelling limitations none of the existing traditional building simulation tools is suitable to meet all the requirements necessary to perform an accurate definition of a complete district in a single simulation model [5]. However, having an integrated model of the district is necessary in order to analyze it as an integrated system optimized for energy efficient operation and peak load reduction.

An offline sequential co-simulation approach has been developed [6], in order to generate the required integrated model for the district. It consists in modelling all the systems and elements included within the boundaries of each of the 3 scales, inside specific models developed in the most suitable tools, and in the definition of the procedures to couple the models, taking into account the dynamics present in the interfaces between component, building and district level models.

A key issue in relation to the implementation of the co-simulation approach is to establish the scope and boundaries of each of the models with very clear interfaces, where interaction dynamics are well known and can be accurately modeled. EnergyPlus has been selected for building and building system modeling and TRNSYS for cooling collector and thermal network modeling. Figure 5 displays the information data flows between the involved models.

In order to evaluate the potential of the SDTN, the co-simulation procedure has been applied to a virtual district heating system defined according to the following criteria:

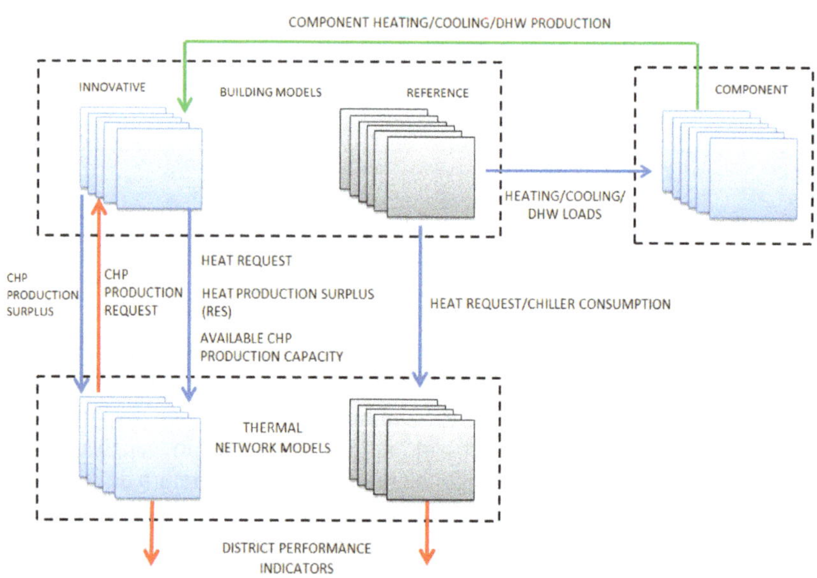

Figure 5: Co-simulation procedure architecture.

- District formed by ten buildings (2 office, 5 educational, a dormitory and 2 hotel buildings), connected to a two pipe distribution loop according to a typical topology, and with an acceptable balance between buildings that act as net consumers and buildings that act as net producers.
- District plant, formed by a CHP unit and two natural gas boilers.
- Absence of solar collectors in the reference scenario.
- Innovative sorption collector deployment on buildings with high DHW demands and conventional solar collector deployment on the rest of the buildings.
- Deployment of conventional absorption chillers on all the buildings of the district.

Starting from the reference scenario, the implementation of the innovative functionalities (bidirectional heat exchange and dual nature of the network) takes place sequentially, in order to have the possibility to evaluate the impact of each of them in a separate and a combined way. Following this approach, the obtained results can be linked with the defined thermal network typologies to estimate their specific potential (Smart Thermal Network, Dual Thermal Network and the Smart Dual Thermal Network).

According to this procedure the potential of these thermal network typologies has been analysed for 2 different locations (Madrid and Bilbao), in order to include the impact of climatic conditions in the performed analysis. Figure 6 displays the EnergyPlus model of one of the buildings of the district. The usage patterns, included in this models have been adjusted to reproduce the specificities of the existing activities and the impact of user behavior on the energy requested by the buildings (occupancy patterns, internal gains, ventilation rates, thermal/visual comfort profiles, internal air quality and DHW demand profiles). Additionally, the geometry of the buildings provides high availability of roof or façade surface for solar collector deployment, with moderate shading and the possibility to deploy solar collectors with acceptable values of tilt and orientation.

On the other hand, thermal envelope quality (insulation, thermal mass, and airtightness) has been adjusted according to the considered climatic boundary conditions and current good practice standards, in order to have realistic values for the demands of the buildings. Table 5 displays the specifications of the Heating, Ventilation and Air Conditioning (HVAC) and distributed generation systems deployed at building level.

Figure 7 displays, the scheme of the TRNSYS model including the district plant, the distribution network and the connected buildings. The considered modelling detail enables the evaluation of all the existing district levels dynamics (heat generator performance, physical location of the buildings inside the district, distribution losses, pumping consumption, thermal substation efficiency and Virtual District Plant management).

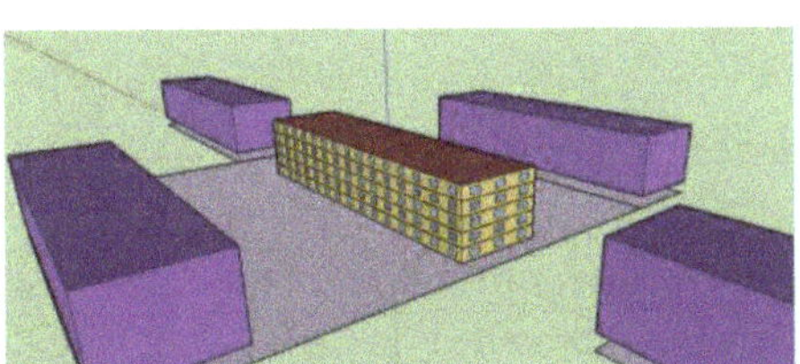

Figure 6: Geometry of the office buildings of the district.

Table 5: HVAC and distributed generation systems deployed at building level.

Reference scenario	
Heat supply	Connection to the thermal network
Cooling	Local through conventional compression chillers
Building connection	Unidirectional
HVAC system typology	Centralized system with 4 pipe distribution
Terminal units	4 pipe fancoils
Ventilation	AHU-s with preheating/precooling hydronic coils, heat recovery, demand controlled ventilation and free cooling
Settings	Heating system supply/return temperature: 60/50°C Cool water supply/return temperature: 7/12°C DHW production and storage temperature: 60°C
Innovative scenarios	
Heat supply	Connection to the thermal network, solar collectors and Virtual District Plant (Capstone C30)
Cooling	Conventional single effect absorption chillers (Heat supply from the thermal network) and innovative cooling collectors
Building connection	Bidirectional
HVAC system typology	As reference scenario
Terminal units	As reference scenario
Ventilation	As reference scenario
Settings	Conventional absorption chiller source temperature: 85°C Cooling collector heat source temperature: 90–95°C

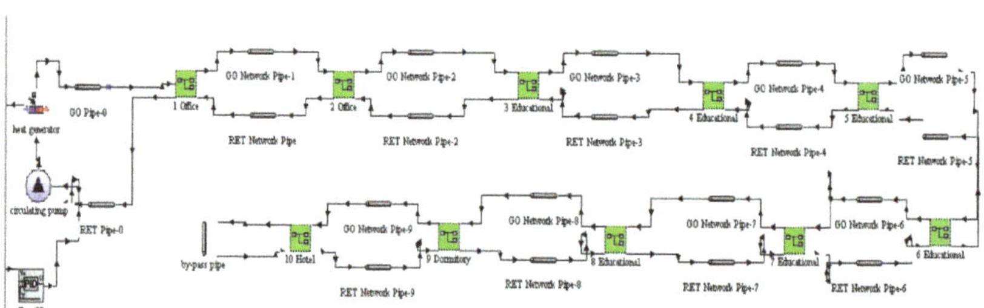

Figure 7: TRNSYS district heating system model.

5 RESULTS AND DISCUSSSION

Tables 6 and 7, summarize the main results obtained for Bilbao and for Madrid:

In the following lines, the results provided by the network potential evaluation process are summarized.

Table 6: Innovative thermal network potential assessment (Bilbao).

KPI-s	Reference Scenario	Smart TN	Smart Dual TN	Dual TN
Primary energy cons. (MWh)	3551.92	2987.30	3349.20	3577.70
Cooling electricity cons. (MWh)	275.2	255.5	34.25	34.25
Plant electricity cons. (MWh)	183.5	224.5	245.2	245.2
Electricity prod. (MWh)	1016.80	745.54	1143.40	1228.40
Operational cost savings (%)		11.87	12.40	6.96
Solar prod. increase (%)		62.77	200.67	134.45
Electricity prod. increase (%)		–26.67	12.45	20.81
Virtual Plant prod. increase (%)			84.44	

Table 7: Innovative thermal network potential assessment (Madrid).

KPI-s	Reference Scenario	Smart TN	Smart Dual TN	Dual TN
Primary energy cons. (MWh)	4412.02	3491.25	3731.00	4281.61
Cooling electricity cons. (MWh)	626.3	551.3	23.1	23.1
Plant electricity cons. (MWh)	135.4	162	169	169
Electricity prod. (MWh)	1033.09	705.92	1309.50	1461.90
Operational cost savings (%)		17.55	28.67	17.96
Solar prod. increase (%)		37.05	224.69	163.77
Electricity prod. increase (%)		–31.76	26.76	41.51
Virtual Plant prod. increase (%)			45.28	

- All the defined thermal network typologies produce significant operational cost savings, but the highest potential is displayed by the Smart Dual Thermal Network, with savings of up to 28.67% (Madrid). However, it is believed that the performed analysis underestimated the existing potential, due to a significant oversizing of the CHP capacity considered for the reference scenario.
- The Smart Network can provide an increase of the solar production of up to 62% (Bilbao).
- The Dual Network can provide an increase of 163.77% of the solar production and an increase of the total electricity production of up to 41.51% (Madrid).
- The Smart Dual Thermal Network can provide an increase of the solar production of up to 224.6%, an increase of the electricity production of the Virtual Plant of up to 45.28%, and a total electricity production increase of up to 26.76% (Madrid).

According to the performed analysis the impact of the innovative functionalities of the thermal network are maximized in locations with high solar irradiation availability and the presence of high cooling demands in the buildings of the district. In any case, it is necessary to extent the performed assessment to further study the impact on the potential of the Smart Dual Thermal Network of climatic conditions, the evolution of absorption technologies, feed in tariffs and energy prices.

6 CONCLUSIONS

Upgrading traditional district heating systems to optimized heating and cooling networks, operated as integrated systems, with optimized RES and CHP contribution must face huge technical and economic obstacles, such as the construction of a district cooling plant and the deployment of a dedicated cooling distribution network.

The innovative Smart Dual Thermal Network concept based on RES, CHP, and locally deployed sorption solutions, as generation technologies, has been developed in the frame of the FP7 A2PBEER project, to overcome these limitations, enabling the transition of buildings from consumers to prosumers. Additionally, in order to enable the configuration of the Virtual District Plant necessary for the implementation of the thermal network, and to facilitate the smart interaction between the buildings and the network, the Smart Dual Thermal Substation has been defined.

Finally, the first assessment of the potential of this innovative thermal network, through a specific co-simulation procedure based on EnergyPlus building models and a TRNSYS district heating system model, has been presented with promising results for 2 different climatic conditions.

ACKNOWLEDGEMENTS

The research activities leading to the described developments and results, were funded by the FP7 project A2PBEER, under grant agreement No 906090. Special thanks to Olof Hallström and ClimateWell AB for making the TRNSYS model of the innovative sorption system and developing the component level simulation work.

REFERENCES

[1] Hallström, O., Design optimization of a sorption integrated sydney type vacuum tube collector. *Journal of Solar Energy Engineering,* **139**(2), p. 021007, 2016. https://doi.org/10.1115/1.4034912

[2] Sterling, R., Giretti, A., Bonvini, M., O'Neill, Z., Wetter, M., Cavey, M.V., Costa1, A., Boehme, G., Zuo, W., Klein, R., Dong, B. & Keane, M.M., *IEA Annex 60 activity 2.3: model use during operation, approach and case studies*. Proceedings of the 14th Conference of International Building Performance Simulation Association, Hydrabad, India, 2015.

[3] Euroheat & Power: Guidelines for District Heating Substations, 2008. Available at: www.euroheat.org/documents/Guidelines District Heating Substations.pdf. (accessed 2013).

[4] Johansson, P.O., Lauenburg, P. & Wollerstrand, J., *Improved cooling of district heating water in substations by using alternative connection schemes*. Proceedings of the 22nd International Conference on Efficiency, Cost, Optimization, Simulation and Environmental Impact of Energy Systems, Parana, Brazil, 2009.

[5] Hensen, J.L.M. & Lamberts, R. (eds), *Building Performance Simulation for Design and Operation,* Routledge, UK, ISBN: 978-0-415-47414-6, 2011.

[6] Fuchs, M., Dixius,T., Teichmann, J., Lauster, M., Streblow, R. & Müller, D., *Evaluation of interactions between buildings and district heating networks.* Proceedings of the 13th Conference of International Building Performance Simulation Association, Chambery, France, 2013.

WIND ASSESSMENT FOR MICRO WIND TURBINES IN AN URBAN ENVIRONMENT

SIMON VAN OVEREEM, LOUIS ALEN, YAIR BROUWER, ANDRE D. VAN DAM, GLENN M. VAN DEKKEN, GEOFFREY H. GARRETT, SVEN GEBOERS, JELLE A. W. POLAND, HARRY W. S. ALDRIDGE, VINIT V. DIGHE & NIKOLAOS CHRYSOCHOIDIS-ANTSOS
Delft University of Technology, The Netherlands.

ABSTRACT

Wind flow in urban environments could be seen as a potential source of energy. This form of energy could be exploited by means of micro wind turbines placed along the existing infrastructures. To test this, an outdoor campaign was organised, which recorded the wind characteristics at different locations around a highway noise barrier in Delft, the Netherlands. The real-time data set was validated with a two-dimensional Computational Fluid Dynamics study. Both the influence of the high turbulence and the inflow angle on the positioning of the micro wind turbines are assessed for the case of perpendicular flow towards the plane of the noise barrier. Results indicated that integrating micro wind turbines with the noise barriers proves advantageous due to the flow velocity increment downstream. Lastly, a noise assessment was conducted in order to determine the optimal spacing between micro wind turbines, which impacts its social acceptance.

Keywords: CFD, noise assessment, noise barrier, urban environment, wind turbines

1 INTRODUCTION

The impetus to sustainability and advances in technology has shown a constant development in renewable methods for generating energy [1]. Wind energy is an inexhaustible form of energy that is already being harnessed on a large scale, both offshore and onshore [2]. Additionally, high concentrations of wind flow in urban environments could be seen as a potential form of energy [3]. Studies indicate that this unrealised form of energy could be exploited by means of micro wind turbines (MWTs) mounted on existing infrastructures, i.e. the noise barriers framework, from hereon referred to as NB. Results indicate that the NB structures lead to an increase in wind speed downstream and thus provide a viable position for the MWTs [4]. Integrating MWTs into the existing infrastructures could also reduce the overall costs involved in setting up the turbine, and provide power for local applications like street lighting and electrical vehicle charging in the future. This study investigates the integration of MWTs in the urban framework. More specifically, a wind site assessment study for wind turbine application is presented.

Accurate measuring techniques to measure turbulent wind characteristics for wind harvesting in urban environments have only recently been developed and experimented upon [4]. Insights into the possible applications of MWTs in an urban environment demand a site-specific study, which was achievable through an outdoor measurement campaign. The assessment was further enhanced by combining the measurements with Computational Fluid Dynamics (CFD) simulations to allow the representation of the flow around the NB. A significant concern is that the turbulent nature and the inflow angle of the wind will negatively affect the efficiency of the turbine, and thereby affecting the turbine's overall performance [5].

The operation of micro wind turbines at a community level requires standardisation and social acceptance. NBs have been positioned in order to reduce noise levels near residential areas caused by vehicles and other disturbances along highways [6]. The level of noise produced by wind turbines is a contentious issue and could potentially lead to opposition from local communities.

DOI: 10.2495/EQ-V2-N4-327-338

Therefore, a noise assessment is performed to ensure that the noise created by MWTs does not exceed the local maximum sound level, which is 41 dB by night and 47 dB during the day [7].

2 METHODOLOGY

The following section presents the description of the test-setup for the outdoor experimental campaign. Also, a detailed description of the CFD methodology and the empirical calculations for the reference wind turbine's performance are reported in this section.

2.1 Campaign Site

An outdoor measurement campaign was conducted at the intersection of the provincial road N470 and the A13 highway in Delft, the Netherlands. The test-setup was positioned on an elevated part, next to the 5.20 m tall NB assemblage, which is located 10 m away from the asphalt of the road. Eight WindMaster Pro anemometers (WM) [8] were used for the measurements, which were distributed over three poles and installed at different heights. The test-setup and a section cut are shown in Fig. 1.

The anemometers record three-dimensional wind speed data (x-, y- and z-directions) and can measure wind velocities up to 65 m/s [8]. Based on the readings, the wind direction in the horizontal plane $(\overline{\gamma}+\psi)$ and the vertical inflow angle (θ) can be defined as shown in Fig. 2. The yaw angle (ψ) is then defined as the difference between the mean wind direction for the two hour data set $(\overline{\gamma})$ and the present wind direction $(\overline{\gamma}+\psi)$.

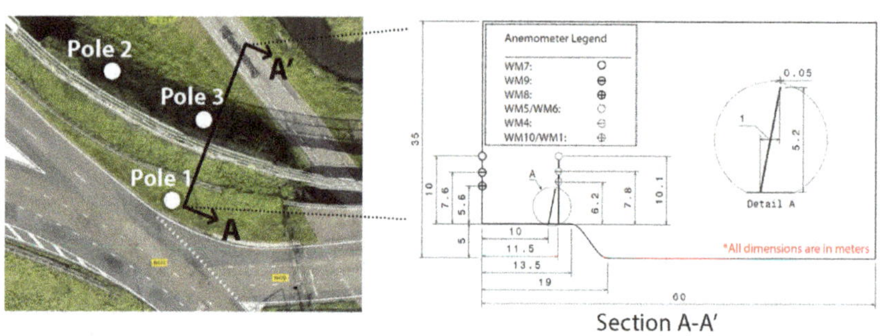

Figure 1: Aerial image of the test-setup (left) and the section cut indicating the measurements (right).

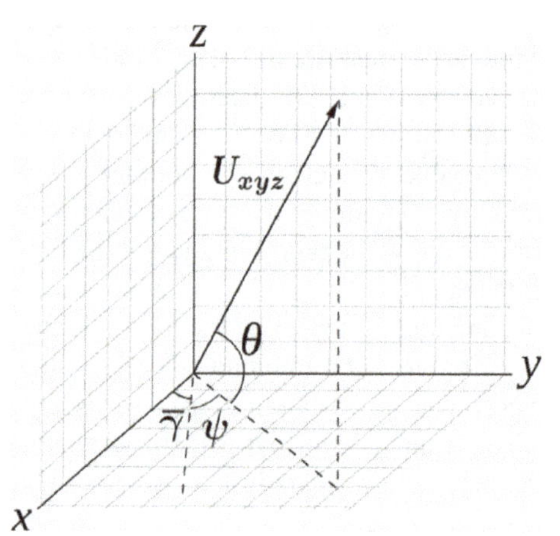

Figure 2: Coordinate system with velocity vector (U_{xyz}) and relevant angles.

The data acquired by the anemometers was collected at a sampling frequency of 4 Hz over a time interval of two hours. A 3-sigma outlier removal method was used such that every data point outside the range $\overline{U}\pm3\sigma$ was deleted, with U and σ being the mean and standard deviation of each data set, respectively. For the wind turbine performance analysis, the velocity component of interest is U_{xy}. For every anemometer, the U_x, U_y and U_z component were recorded and therefore the U_{xy} velocity component, yaw angle and the turbulence intensity was calculated using a 10 minutes sampling period. The turbulence intensity was calculated using the *xyz* velocity vectors.

2.2 CFD Model

The numerical flow is governed by the continuity and the Navier-Stokes (in the form of momentum conservation) equations (see eqn. (1) and eqn. (2)). A k-ω SST (shear stress transport) was used as a turbulence model. Preliminary investigations have shown that the model is suitable for cases involving adverse pressure gradients [9].

$$\frac{\partial u_i}{\partial x_i} = 0 \tag{1}$$

$$\rho\frac{\partial u_i}{\partial t} + \rho\frac{\partial(u_i u_j)}{\partial x_j} = -\frac{\partial p}{\partial x_i} + \mu\left(\frac{\partial^2 u_i}{\partial x_j^2} + \frac{\partial^2 u_j}{\partial x_j \partial x_i}\right) \tag{2}$$

The computational domain used for the numerical study is shown in Fig. 3. The domain is discretised using a structured mesh. Bi-geometric bunching law is applied to achieve a finer mesh near the NB geometry, and gradually coarsening as it moves further upstream and further downstream.

The convergence towards a unique numerical solution is heavily dependent on the implementation of correct boundary conditions suited to the real geometry being modelled. Velocity inlet and pressure outlet boundaries are defined to simulate the far-field flow conditions, as shown in Fig. 3. A quadratic velocity profile is described, which is obtained through the real measurements of the anemometers placed upstream of the NB. This allows a realistic flow

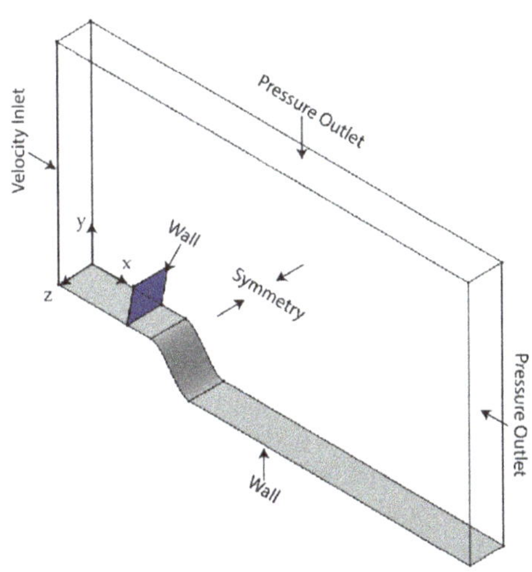

Figure 3: Flow domain designed within ANSYS® ICEM with associated boundary conditions.

Table 1: Organisation of anemometers into Set 1 & 2, before and after the NB respectively.

Set 1	Pole 1	WM7	WM9	WM8
Set 2	Pole 2	WM1	-	WM6
	Pole 3	WM10	WM4	WM5

Table 2: Boundary conditions and values of the flow domain in the CFD model.

Boundary	Type	Condition
$yz_{x=0}$	Velocity Inlet	V=$U_{x=0}$(y) I=30.67%
$yz_{x=60}$	Pressure Outlet	P=101325 Pa I=24.98%
$xz_{y=30}$	Pressure Outlet	P=101325 Pa I=24.98%
Surfaces	Wall	V=0
All xy-planes	Symmetry	w.r.t. boundary

Table 3: Mass flux distribution for the numerical solution.

Boundary	$xz_{y=30}$	$yz_{x=0}$	$yz_{x=60}$	Sum
Mass Flux [kg/s]	-67.657	142.497	-74.839	1.0

field initialisation and improves the solution approximation of the flow downstream, which is of primary interest. The turbulence intensity (I) is set to the average turbulence intensity for Set 1 & 2, which is obtained from the real measurements. The sets and their respective anemometers can be found in Table 1. As the model generated is represented by a two-dimensional profile, the side wall boundary conditions are described by symmetry within a unit distance of one another.

The solutions of momentum, turbulent kinetic energy and specific dissipation rate were second order accurate based on upwind differential scheme. The boundary conditions used for the CFD model are displayed in Table 2.

In order to verify the completeness of the CFD solution, the mass flux of the system is calculated. Table 3 indicates the sum of the mass flux within the system. It is notable that there is 1 kg/s of mass flow that is unaccounted for in the residuals. This is close to negligible compared to the entire mass flow of the system which indicates substantial verification of the system when consideration is placed on the relative value. Therefore, the solution is verified.

2.3 Reference Turbine

The reference turbine used for the analysis is a 1 kW MWT designed by FuturEnergy with a cut-in wind speed of 3 m/s. For the reference case, the power-velocity curve given by the manufacturer is used to compare with the deduced performance under outdoor conditions.

The power output of the wind turbine is given by eqn. (3). The power generated by a wind turbine (P) is dependent on the power coefficient of the wind turbine (C_p), the air density (ρ), the area of the rotor (A_d) and the free stream velocity (U_∞) [10, 11].

$$P = \frac{1}{2} C_p \cos^3(\psi) \rho \, A_d \, U_\infty^3 \tag{3}$$

Note that for a yawed flow, the maximum power is proportional to $\cos^3(\psi)$. The yaw angle has therefore a direct influence on the power generated [5, 12].

2.4 Sound Assessment

Integrating MWTs along highway NBs introduces a new source of sound, which may emit intolerable sound levels into the nearby community. Therefore an optimal spacing has to be determined to ensure that the MWTs do not exceed the allowable 41 dB sound level. Apart from the horizontal wind velocity, the sound level at the nearest facade also depends on the spacing (b) between the MWTs on the NB as well as its distance (d) to the nearest facade, as illustrated in Fig. 4. Note that every MWT will be treated as an individual sound source. A relation between the sound power produced and the maximum wind velocity was found by fitting data provided by FuturEnergy, whom have performed a noise assessment on their product [13]. This relation is given by eqn. (4), which gives the maximum sound power (P_{SW}) in microwatt as function of the maximum wind velocity (V_{max}).

$$P_{SW} = 8.04e^{0.2216 V_{max}} \tag{4}$$

It is necessary to take both the sound produced by the highway traffic as well as the sound produced by the MWTs into account. The NB influences the propagation of the sound waves emitted by the highway. Explaining the working principles of a NB in depth, however, goes beyond the scope of this report. The NB lowers the sound level by a certain number (which is typically in the range of 10–40 dB [6, 14]). This number is dependent on the material transmission loss factor. By reducing the sound level, the NB effectively lowers the highway noise power which in turn reduces the noise level at the houses. For calculations a sound level of 53.5 dB was assumed [6, 15], this is at a distance of 10 m from the highway and includes the loss due to the NB.

Wind gusts will only give a temporary high horizontal wind velocity, therefore a 3-sigma confidence interval was used such that the maximum horizontal wind velocity that has to be taken into account is $V_{max} = \overline{U} + 3\sigma$ With the assumption of a spherical sound distribution, using an infinite series (eqn. (5)) to calculate the contribution of each individual MWT at a point location (Fig. 4), the total sound intensity at the facade is calculated, as stated in eqn. (6).

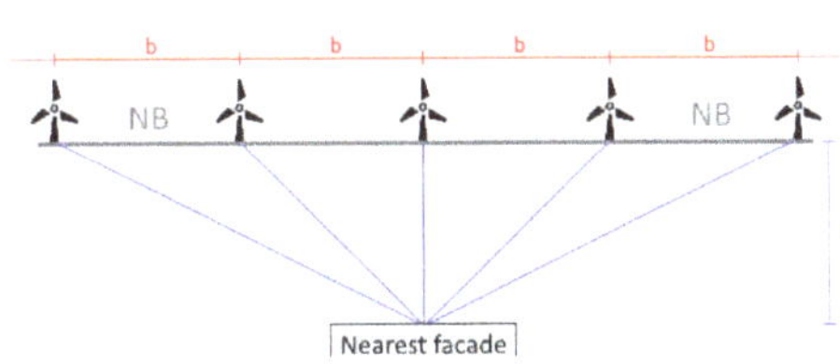

Figure 4: Spacing between the MWTs (b) and the distance of the NB to the nearest facade (d).

$$\sum_{n=1}^{\infty} \frac{1}{n^2 + a} = \frac{-1 + \sqrt{a} \cdot \pi \cdot \coth\left(\sqrt{a} \cdot \pi\right)}{2a} \tag{5}$$

$$I_w = \frac{P_{SW}}{4\pi}\left(\frac{1}{d^2} + 2 \cdot \sum_{n=1}^{\infty} \frac{1}{(\mathrm{n} \cdot \mathrm{b})^2 + d^2}\right) = \frac{P_{SW}}{4} \frac{\coth\left(\frac{d \cdot \pi}{b}\right)}{d \cdot b} \tag{6}$$

3 RESULTS

This section initially introduces the mean flow properties after which the CFD model is validated. Using the results, the annual energy production for the reference turbine placed at every anemometer location height in the test-setup was calculated. Finally, results pertaining to the noise assessment and economic impact are described.

3.1 Mean Flow Properties

Table 4 shows the position, mean wind direction (MWD), loss due to yaw, turbulence intensity (I), mean value of the horizontal velocity component (U_{xy}) and the mean value of the velocity component in the z-direction (U_z) for each anemometer. By comparing the average loss factor due to yaw angle at different heights, it can be concluded that at a height of approximately 6 m, the loss factor is lower behind the NB compared to the free stream. As the height increases, the loss factor behind the NB decreases, while the loss factor of the pole in the free stream (pole 1) increases. The average turbulence intensity values before and after the NB are 31% and 24%, respectively. The higher turbulence intensity before the NB could be due to traffic passing by. Furthermore, it can be observed that U_{xy} decreases with increasing height in the free stream, while for the poles behind the NB, it increases with increasing height. Conversely, the U_z velocity component increases with height in the free stream, but decreases with height after the NB. Finally, a noteworthy feature result can be observed for WM10, where the mean wind direction is equal to 1.32°. This appears to be the result of the sample angles being placed almost perfectly symmetrical around 0°. This can also be concluded from the fact that its yaw loss factor is smaller than the others.

Table 4: Mean flow properties recorded for different anemometers.

Anemometer (Pole #)	**MWD [deg]**	**Yaw loss [-]**	**I [%]**	U_{xy} **[m/s]**	U_z **[m/s]**
WM 8 (1)	-12.81	0.82	34	3.50	0.33
WM 9 (1)	-20.80	0.90	28	4.16	0.45
WM 7 (1)	-13.00	0.88	30	3.86	0.55
WM 1 (2)	-23.75	0.78	28	3.97	1.57
WM 6 (2)	-22.45	0.92	25	5.02	0.92
WM 10 (3)	1.32	0.75	24	4.08	1.70
WM 4 (3)	-18.74	0.86	24	4.41	1.13
WM 5 (3)	-15.50	0.90	23	4.73	0.87

3.2 CFD Analysis Results

The solution to the CFD model was considered converged after residuals fell below the order of 10^{-5} which provides significant accuracy regarding the discretisation, iteration and round off error in the numerical model. Figure 5 shows that the locations where the anemometers are located are not in the highest magnitude of flow potential.

Comparison of the velocity profiles before and after the NB in Fig. 6a lead to two observations. Firstly, the velocity magnitude before the NB drops closely to zero at collision with the NB. Secondly, the velocity magnitude at Set 2 shows significant increase, predominantly at a greater height.

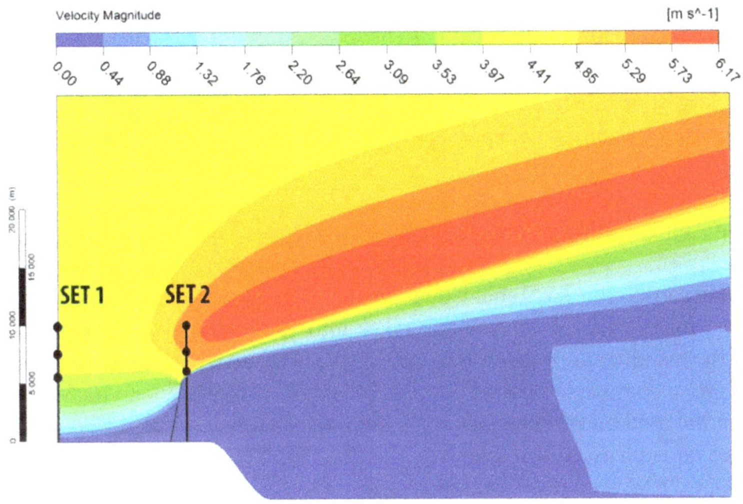

Figure 5: Velocity magnitude contours plotted using ANSYS® Fluent showing anemometer sets. The right-most red color of the legend represents the greatest magnitude in velocity and the left-most blue color, the lowest.

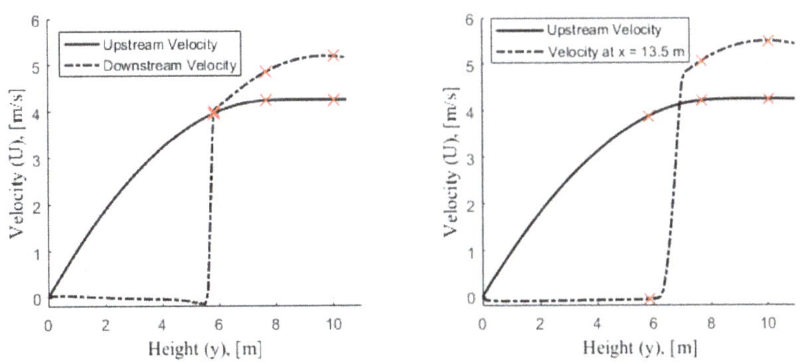

(a) Velocity profile upstream of the NB (Set 1) and downstream from the NB (Set 2). The crosses (×) depict the anemometer locations through which interpolation provided the inlet velocity profile $U_{x=0}(y)$.

(b) Velocity profile upstream of the NB at an x-coordinate of 13.5 m depicting a location of greater wind velocities than that of the location of Set 2 (Pole 2 & 3). For comparative purposes, upstream velocity is shown.

Figure 6: Wind velocity analyses on the CFD model from the velocity contours in Fig. 5. The crosses (×) depict the anemometer locations.

Table 5: Absolute and model errors between the CFD analysis and the outdoor measurements.

		Physical [m/s]		Model error [%]	
Location	**Model [m/s]**	Pole 2	Pole 3	Pole 2	Pole 3
WM5/WM6	5.182	5.06	4.77	2.35	7.95
WM4	4.893	-	4.43	-	9.46
WM10/WM1	4.167	3.98	4.10	4.49	1.61

3.2.1 Validation of CFD Model

Values deduced from CFD are compared to the values of Set 2 of the anemometers providing an estimation of the model uncertainty. Table 5 shows that the model errors fall below 10%, wherein, the largest error occurs for WM4 at pole 3 (9.46%).

With the limited amount of data points available in the data sets, the level of model error is considered reasonable.

3.3 Potential Annual Energy Production (AEP)

As discussed in subsection 3.1, Table 4 displays the yaw loss for every anemometer. To obtain a power curve for every anemometer that takes into account the yaw loss, the power curve provided by the manufacturer (theoretical ceiling) [11] has to be multiplied with the yaw loss. This results into a comparison power curve for every location. The one for WM6 is given in Fig. 7. Using the method of moments, a Weibull distribution for the different anemometers was plotted to obtain a probability density function [16, 17].

Multiplying each Weibull distribution with the corresponding power curve under yaw conditions results in a graph, which upon integration gives the expected power output. From this value, the AEP can be obtained in kWh. Note that this annual energy estimation is still theoretical as it does not take turbulence intensity into account. If the turbulence intensity is higher than 21% on a ten minute interval, the theoretical power will be increased by 4% [18]. From subsection 3.1, it can be seen that the turbulence intensity is higher than 21% at every location, and therefore an increase in power of 4% has to be taken into account when estimating the real AEP.

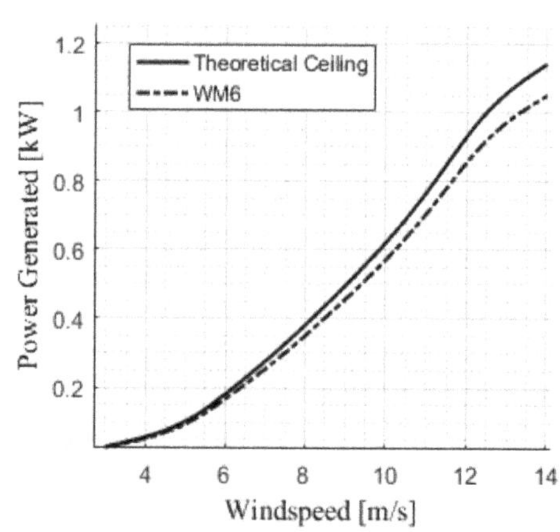

Figure 7: Power produced as a function of wind speed for the position of WM6.

Table 6: Theoretical AEP for different anemometers.

Windmaster	WM1	WM4	WM5	WM6	WM7	WM8	WM9	WM10
Energy [kWh]	450.2	650.1	827.4	1024.1	458.9	339.2	588.3	454.2

The final AEP for every anemometer location has been displayed at Table 6. It can be deduced that the produced power decreases with height for the free stream pole (pole 1). For the two poles behind the NB however, the power generated increases with increasing height. From Table 6, it can be deduced that the maximum power output occurs at the position where WM6 is located. However, from Fig. 5, it can be seen that WM6 is not the place where a maximum wind velocity occurs. A maximum wind velocity of 6.2 m/s can be obtained 22 m behind the noise barrier. Using this mean velocity, a resulting AEP of 1553.1 kWh was found.

3.4 Socio-Economic Impact

The socio-economic impact will first give the effects on the sound level due to the MWTs. Not only the sound level is an important concern, but also the feasibility of the implementation of MWTs. Therefore, the economic perspective of integrating MWTs in the urban framework will be considered.

3.4.1 Sound Assessment

From subsection 2.4, a horizontal wind velocity of 9.25 m/s was found for WM6. Using eqn. (4), this results in a turbine sound power of 0.0625 mW. Furthermore, the sound power of the highway can be calculated to be 0.2813 mW. The total intensity can then be calculated by adding the sound intensity due to the MWT (I_W) and the sound intensity due to the highway (I_H). Given a horizontal wind velocity of 9.25 m/s, a contour plot of the sound level based on the spacing of the wind turbines and the distance to the NB can be produced. The contour plot is shown in Fig. 8.

For the test location in Delft, the distance to the nearest facade is 40 m. This results in a minimal spacing of 100 m to stay under the noise level limit. If the spacing is lower than this minimal spacing, the allowed noise level is surpassed.

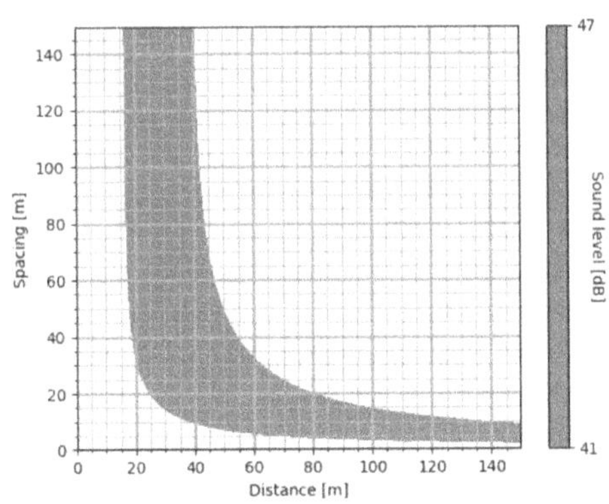

Figure 8: Contour plot providing sound levels, levels of interest (47 dB (left) and 41 dB (right)) are accentuated. Sound levels are in function of the spacing b and the distance d.

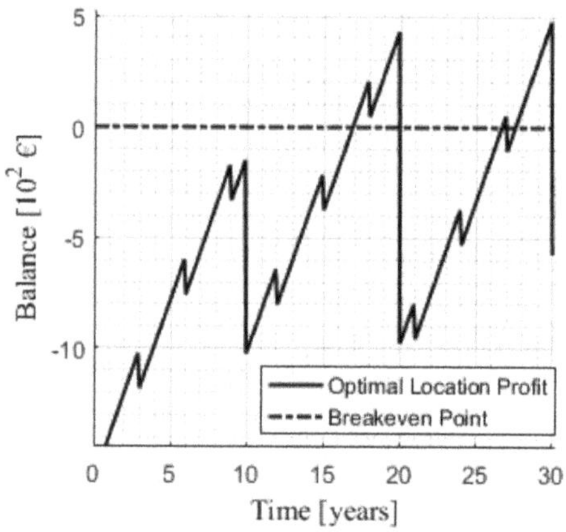

Figure 9: Operating income generated incorporating maintenance costs per MWT (in EUR).

3.4.2 Economic Impact

In order to make the installation of MWTs an economically viable plan, it needs to be confirmed that the power produced by these micro wind turbines generates sufficient income to amortise the purchase and maintenance costs of the MWTs. To visualise the economic benefit of the MWTs, Fig. 9 shows the operating income of the MWTs as a function of time. Note that this is calculated for the best-case scenario, where the MWTs are placed at a region of maximum velocity obtained from the CFD model, being 6.2 m/s, at a distance of 22 m from the noise barrier (see Fig. 5).

The assumption is made that the energy produced can be directly used, saving the costs of buying electricity at the price of 0.18190 EUR/kWh, while estimated electricity network costs (to put electricity on the net) are 0.061 EUR/kWh [19]. The FuturEnergy 1 kW MWT currently costs 1200 GBP (circa 1428 EUR at present). According to article 3.14 of the Dutch civil code for environment, all forms of wind energy devices need a performance and maintenance check at least once a year [7]. Thus, after this initial purchase cost, about every three years a total maintenance cost of 150 GBP (circa 178 EUR) can be expected [20]. Within the range of ten years, the inverter of a MWT also needs replacement, which has a cost estimated to be 750 GBP (circa 893 EUR). At the end of its lifespan, about 20 years, the MWTs have to be replaced [21]. It can be seen that the turbine starts to generate a net revenue after 17 years. Note that the costs are taken very conservatively.

4 CONCLUSION

From the analysis of the data set, it can be concluded that the best position for placing a micro wind turbine (MWT) is at a height of 10 m downstream of the noise barrier (NB), which corresponds to WindMaster 6. At this position, the annual energy production (AEP) of the MWT is estimated to be 1024.08 kWh. However, further downstream of the NB assemblage a region of even greater flow potential is obtained. This is done using the validated Computational Fluid Dynamics (CFD) model. The AEP at this position will then be 1553.1 kWh. It is therefore apparent that the optimal position for placing a MWT will be within this region.

As NBs are primarily designed to reduce the noise emitted by highway traffic, adding another source of sound (the MWT) on top of the NBs is not beneficial. Using data provided by the manufacturer and the data set obtained from the measurements, a minimum spacing of

100 m between two MWTs was found to ensure that the legal sound limit of 41 dB will not be surpassed at the nearest house. Regarding economic sustainability, the current estimations for power output at the optimal location according to the CFD model outweigh the cost estimates for the turbine upkeep and maintenance cost provided by the manufacturer after 17 years. Note that these costs are taken conservatively. The wind turbine would generate revenue sooner, especially when MWTs are deployed on a large scale. Ultimately, this report has demonstrated that the flow over a NB increases its velocity, which in turn increases the power output of a MWT placed behind it. The research has shown that the implementation of MWTs in an urban environment is a feasible future energy solution.

5 RECOMMENDATIONS

With the goal to further improve on this research, a few additions that were considered to be added to this article will be listed in this section. These considerations were beyond the scope of the original question answered and could therefore lead to potential topics for further research.

The effect of the vertical inflow angle on the MWT's performance was ignored. The vertical inflow angle and the wind directivity will realistically cause a fluctuation in the energy output of the MWT; the significance of the performance fluctuation could be investigated by using a 3D CFD model over the currently used 2D CFD model.

Furthermore, the cost analysis performed in this study was executed using conservative values for maintenance and inverter replacements. One could consider doing an experiment in which a MWT is placed in similar conditions and record the costs that are required for one MWT in operating conditions throughout one year. A detailed cost assessment of various aspects involved would give a better view on the economical viability.

Finally, in this study, only two hours of measurements were taken due to facility limitations. By using longer measurements throughout the year, it is possible to include measurements through all weather conditions. Generating an annual energy production based on data recorded over the span of an entire year, would allow for better insight into the effectiveness and feasibility of MWTs in an urban environment.

REFERENCES

[1] Dincer, I., Renewable energy and sustainable development: a crucial review. *Renewable and Sustainable Energy Reviews*, **4**, pp. 157–175, 2000. https://doi.org/10.1016/s1364-0321(99)00011-8

[2] Tong, W., Wind Power Generation and Wind Turbine Design, *WIT Transactions on State-of-the-art in Science and Engineering*, vol. 44, WIT Press: Southampton and Boston, p. 20, 2010.

[3] Britter, R.E. & Hanna, S.R., Flow and dispersion in urban areas. *Annual Review of Fluid Mechanics*, **35**, pp. 469–496, 2003. https://doi.org/10.1146/annurev.fluid.35.101101.161147

[4] Abohela, I., Neveen, H. & Dudek, S. Effect of roof shape, wind direction, building height and urban configuration on the energy yield and positioning of roof mounted wind turbines. *Renewable Energy*, **50**, pp. 1106–1118 2013. https://doi.org/10.1016/j.renene.2012.08.068

[5] Burton, T., Sharpe, D., Jenkins, N. & Bossanyi, E., *Wind Energy Handbook*, John Wiley & Sons, Chichester, 2001.

[6] Environmental Protection Department Highways Department. *Guidelines on Design of Noise Barriers*, pp. 4–5, 2003.

[7] Dutch Government. Wet- en regelgeving (in Dutch), available from: http://www.wetten.overheid.nl/BWBR0022762/2017-02-02#Hoofdstuk3_Afdeling32_323 (Accessed on: 31 April 2017).
[8] Gill Instruments Limited. Windmaster Pro 3-Axis Anemometer, available at: http://www.gillinstruments.com/products/anemometer/windmaster- pro.html. (Accessed on: 24 April 2017).
[9] Dighe, V.V., Avallone, F., Tang, J. & van Bussel, G.J.W., Effects of Gurney Flap on the performance of Diffuser Augmented Wind Turbines. *35th Wind Energy Symposium*, American Institute of Aeronautics and Astronautics, p. 3, 2017.
[10] Accon-UK. FuturEnergy: Turbine to scale drawing, available at: http://www.futurenergy.co.uk/ pdf_doc/Airforce1TurbinetoScaleDrawing.pdf. (Accessed on 3 May 2017).
[11] Accon-UK. FuturEnergy: 24 V wind turbine power curve, available at: http://www.futurenergy.co.uk/pdf_doc/24VTurbineSpec&PowerCurve.pdf. (Accessed on: 3 May 2017).
[12] Ozbay, A., Tian, W., Yang, Z. & Hu, H. Interference of Wind Turbines with Different Yaw Angles of the Upstream Wind Turbine, *42nd AIAA Fluid Dynamics Conference and Exhibit*, American Institute for Aeronautics and Astronautics, p. 2719, 2012.
[13] Accon-UK. FuturEnergy: 1kW Airforce Wind Turbine Noise Assessment, available at: http://www.futurenergy.co.uk/pdf_doc/Airforce1NoiseAssessment.pdf. (Accessed on: 27 March 2017).
[14] Vargas, C. Understanding the basics of sound control, April 2007.
[15] Sengpielaudio. Table of sound levels L (loudness of noise) with corresponding sound pressure and sound intensity, available at: http://www.sengpielaudio.com/TableOfSoundPressureLevels.htm. (Accessed on: 28 April 2017).
[16] The Wind Power Program. Wind statistics and the Weibull distribution, available at: http://www.wind-power-program.com/wind_statistics.htm. (Accessed on: 3 May 2017).
[17] Justus, C., Hargraves, W., Mikhail, A. & Graber, D. Methods for estimating wind speed frequency distributions. *Journal of Applied Meteorology*, **17**, pp. 350–353, 1978. https://doi.org/10.1175/1520-0450(1978)017<0350:mfewsf>2.0.co;2
[18] Lubitz, W.D., Impact of ambient turbulence on performance of a small wind turbine. *Renewable Energy*, **61**, pp. 69–73, 2014. https://doi.org/10.1016/j.renene.2012.08.015
[19] U.S. Energy Information Administration. Levelized Cost and Levelized Avoided Cost of New Generation Resources in the Annual Energy Outlook 2017, April 2017.
[20] Energy Saving Trust. Wind Turbines, available at: http://www.energysavingtrust.org.uk/renewable-energy/electricity/wind-turbines. (Accessed on: 1 May 2017).
[21] Accon-UK. FuturEnergy: 1kW Upwind Turbine, available at: http://www.futurenergy.co.uk/turbine.html. (Accessed on: 24 April 2017).

SOLAR WALLS FOR HIGH-PERFORMANCE BUILDINGS

BORISLAV N. STANKOV, NIKOLA G. KALOYANOV & GEORGI D. TOMOV
Technical University of Sofia, Bulgaria.

ABSTRACT

Passive solar design can reduce building energy demand for heating, cooling and ventilation, while also contributing to the comfort, well-being and productivity of the building's occupants. The successful application of passive solar features, such as solar walls, requires a good understanding of the factors influencing their energy performance and a correct assessment of this performance during the design process. This paper discusses some basic design strategies for successful application of solar walls and the factors with the most significant impact on their efficiency. It summarizes the principle results and findings of an experimental study, based on dynamic simulations and test site measurements. The energy performance of various configurations of unvented solar walls was investigated in different climatic conditions. The outcomes of the dynamic simulations were used to develop a simplified quasi-steady-state model, which can be used for approximate evaluation of the heat gains and heat losses through an unvented solar wall on a monthly basis. The model is compatible with the monthly method of EN ISO 13790.

Keywords: EN ISO 13790, energy performance, experimental, green buildings, heat transfer, modelling, passive solar, TRNSYS, Trombe wall.

1 INTRODUCTION

Passive solar design is a coalescence of design decisions concerning basic elements of building architecture, such as shape, orientation and spatial distribution, as well as individual passive systems or elements, which perform specific functions. The building envelope is a crucial element of passive solar design, as it determines the heat transfer processes occurring between the indoor and outdoor environments, including the ingress and distribution of solar radiation in the building. Solar walls can be employed as an element of passive solar design to enhance the capabilities of the building envelope to collect and store solar heat. The accumulated heat can be utilized to reduce the heating demand and/or provide means for natural ventilation.

Solar walls are also commonly referred to as 'Trombe walls' or 'Trombe–Michel walls', after the French engineer Felix Trombe and architect Jacques Michel, who developed a naturally ventilated type of solar wall [1]. A somewhat similar invention has been patented earlier by Morse [2]. Other sources use the terms 'collector-storage wall' or 'thermal storage wall' [3–5]. A solar wall consists of four basic elements: (1) thermal storage mass, i.e. wall; (2) absorber coating on the external surface of the wall; (3) glazing unit installed in front of the external surface of the wall; and (4) air gap between the wall and the glazing unit. The thermal storage mass is usually a solid masonry wall, built from a material which is relatively dense and has a high heat capacity. Examples include concrete, solid (non-porous) bricks, adobe and stone. Liquid-filled containers or encapsulated phase-change substances can be used to achieve greater thermal efficiencies [6, 7]. The absorber coating is a thin layer of material with very high solar absorptivity, e.g. black paint. A high-performance option is to use selective coating which has a very low infrared emissivity, e.g. black chrome [8–11]. Typically, the glazing unit would consist of one or two panes of clear glass, but other transparent materials are also possible. For example, Dowson *et al.* [12] investigate lightweight polycarbonate panels filled with nanoporous aerogel insulation as an alternative.

DOI: 10.2495/EQ-V2-N4-339-351

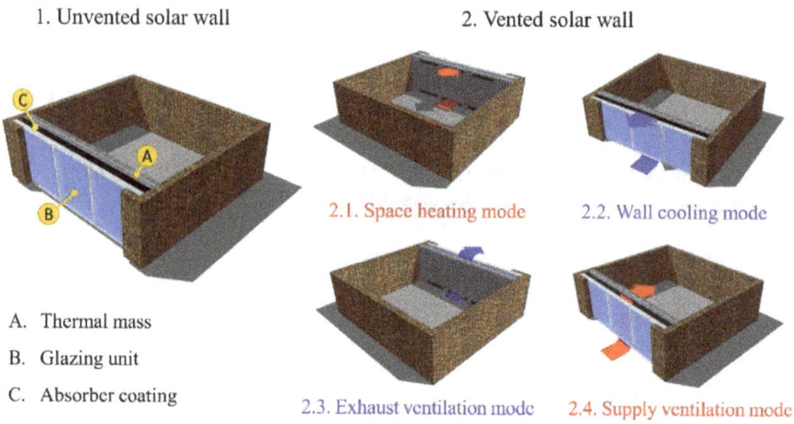

Figure 1: Basic configurations of solar walls.

A conceptual illustration of some basic solar wall configurations is shown in Fig. 1.

In an unvented solar wall heat absorbed at the external surface diffuses through the thermal storage mass and is transferred to the indoor environment with a time lag, which depends on the thermal diffusivity and thickness of the wall. When the solar wall is provided with vents, air can be allowed to flow through the space between the wall and the glazing. The absorption of thermal radiation produces a buoyancy-driven flow (solar chimney effect), whereby cold air entering through the bottom vents rises and gains heat from the absorber, and is subsequently discharged through the top vents. The air velocity and flow rate depend on the temperature differences and the pressure drops occurring through the vents and the air gap. Ventilation can be entirely natural or fan-assisted (if higher and/or constant flow rates are required). Saadatian *et al.* [6] gave a review of various traditional and innovative design concepts and applications of solar walls. Good examples of passive solar design incorporating solar walls can be found in the Zion National Park visitor centre and the NREL visitor centre in the United States [8, 9, 13].

This paper summarizes the principle results and findings of an experimental study on the energy performance of unvented solar walls of the classical type shown in Fig. 1. Therefore, further in this paper, the term 'solar wall' should understood in that context. The experimental study involves numerical analysis, based on dynamic simulations, and physical measurements obtained from a test site.

2 MATERIALS AND METHODS

2.1 Dynamic numerical model

The dynamic model was created in TRNSYS and consists of components from the TRNSYS standard library [14] and the TESS libraries [15], as well as an additional component, programmed for purposes of the experiment [16]. The basic component of the model is Type 36, which contains the method and algorithms for calculation of the heat flows and temperatures in the solar wall. The component is developed by the Solar Energy Laboratory of the University of Wisconsin-Madison [17].

2.1.1 Studied locations and solar wall configurations

Simulations were carried out for 11 340 configurations with variation of the following parameters:

- Thickness (10–50 cm) and material (Table 1) of the thermal storage mass;
- Type of absorber coating (Table 2);
- Type of glazing unit (one or two panes of clear glass);
- Distance between the wall and the glazing (in the range of 2.5–10 cm);
- Surface azimuth (-45°, 0° and +45°);
- Weather data (five locations in Bulgaria).

These simulations comprise the training set, which was used to develop the simplified model. A number of additional simulations were carried out with hourly weather data for 15 different locations in Europe and North America: Valencia, San Francisco, London, Rome, Istanbul, Richmond, New York, Munich, Strasbourg, Boulder, Chicago, Albuquerque, Stockholm and Helsinki. The selected locations are situated in diverse climate zones – ASHRAE climate zones 3 to 7 [18] and zones Csa, Csb, Cfa, Cfb, Dfa, Dfb and BSk, according to the Köppen–Geiger classification [19]. For comparison, four of the locations in Bulgaria (Pleven, Varna, Plovdiv and Sandanski) are situated in ASHRAE climate zone 4 and Köppen–Geiger zone Cfa, while the fifth (Sofia) is situated in ASHRAE zone 5 and Köppen–Geiger zone Cfb. The additional simulations (test set) were used to test and prove the robustness of the developed model and its validity for a wider range of climatic conditions. The test set consists of 4 320 data points.

2.1.2 Simulations and data processing

Each solar wall configuration was simulated twice. In the first instance, it was simulated with no incident solar radiation. In that case heat transfer through the solar wall is driven

Table 1: Properties of thermal storage materials.

Material	Thermal conductivity, [W/(mK)]	Density, [kg/m^3]	Specific heat capacity, [J/(kgK)]
Adobe	0.75	880	1730
Clinker brick	0.85	840	1500
Calcium-silicate brick	1.5	840	2000
Sandstone	1.3	712	2200
Limestone	1.5	720	2180
Concrete	1.7	840	2000

Table 2: Properties of absorber coatings. *(Source: Steven Winter Associates [11])*

Type of coating	Solar absorptivity	Infrared emissivity
Unfinished concrete	0.65	0.87
Flat black paint	0.96	0.87
Selective (black chrome)	0.95	0.11

only by the temperature difference between the indoor and outdoor environments, $Q_{ht,sw}$ (for simplicity, the term further used for $Q_{ht,sw}$ is 'heat losses', although heat flow driven by the indoor-outdoor temperature difference can occur in both directions). The second simulation was used to obtain the actual net heat transfer when solar radiation is taken into account. The solar heat gains through the wall, $Q_{gn,sw}$, were calculated as the difference between the net heat transfer obtained from the second simulation and the heat losses obtained from the first simulation. Equivalently, the net heat transfer between the solar wall and the indoor environment, Q_{sw}, [kWh], is equal to the difference between the heat gains and the heat losses:

$$Q_{sw} = Q_{gn,sw} - Q_{ht,sw}. \quad (1)$$

Both Q_{sw} and $Q_{ht,sw}$ can be positive or negative, while $Q_{gn,sw}$ can only be positive.

2.2 Test site configuration

The test site, shown in Fig. 2, comprises an unvented solar wall integrated in the south façade of a modular house, which is located at the Technical University of Sofia. The solar wall is built of a 12-cm-thick masonry wall with black paint as absorber coating and a clear double glazing, installed at 10 cm from the wall. The wall is divided in two parts – the lower half is made of concrete masonry, while the upper half is made of clay bricks. The test module is equipped with 28 type T thermocouples for measuring the temperature field in the wall and four sensors for measuring the temperatures and air velocities in the air space (Fig. 3). The nearby located weather station provided all necessary weather data, including measurements of total and diffuse solar radiation. Measurements were collected for the heating season of 2014/2015 (22 October–22 April).

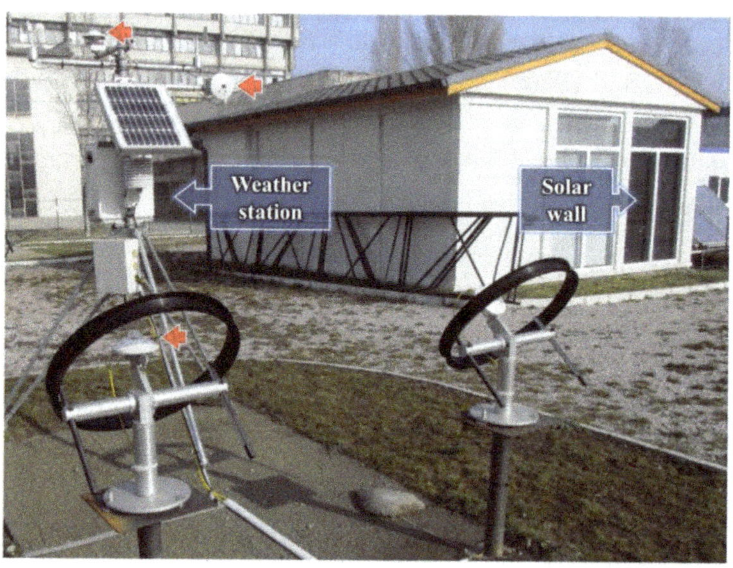

Figure 2: Test site at the Technical University of Sofia.

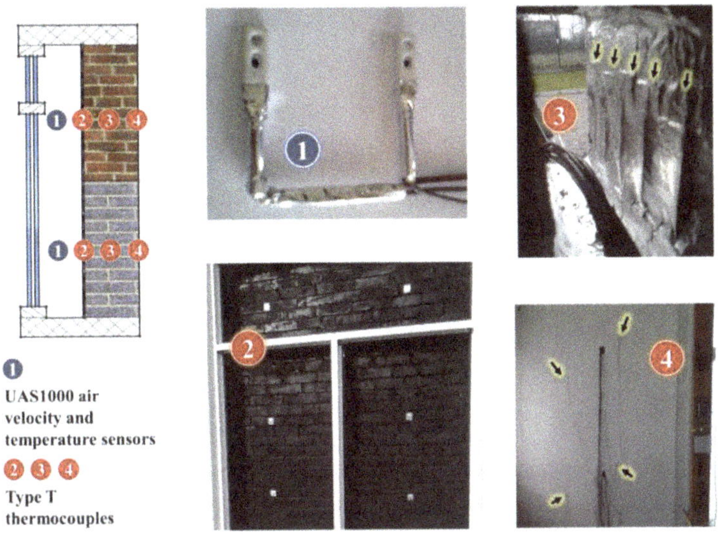

Figure 3: Configuration of the solar wall test module.

3 RESULTS AND DISCUSSION

The outcomes of the simulations were used to develop a simplified quasi-steady-state model, which can be used for approximate evaluation of the heat gains and heat losses through a classical unvented solar wall. The model is compatible with the monthly method of EN ISO 13790 [20]. It is presented here along with a discussion of the basic factors determining the energy performance of the solar wall and its sensitivity to their values.

Heat losses (as defined in section 2.1.2) can be calculated as:

$$Q_{ht,sw} = U_t A_{sw} \left(\theta_{int,set} - \theta_e\right) t / 1000. \tag{2}$$

where: $Q_{ht,sw}$ – heat losses, [kWh]; U_t – overall heat transfer coefficient of the solar wall, [W/(m^{2}K)]; A_{sw} – internal surface area of the wall, [m^2]; $\theta_{int,set}$ – internal set-point temperature, [°C]; θ_e – mean ambient temperature, [°C]; t – duration of the period, [h]. The overall heat transfer coefficient in eqn (2) is used to calculate the theoretical heat losses which will occur in the case of a complete lack of solar radiation (in order to separate the heat losses from the solar heat gains):

$$U_t = 1 / \left(R_{si} + L_p / k_p + 1 / U_e\right). \tag{3}$$

where: R_{si} – thermal resistance of the internal surface, [m^{2}K/W]; L_p – thickness of the wall, [m]; k_p – thermal conductivity of the wall, [W/(mK)]; U_e – external heat loss coefficient, [W/(m^{2}K)]. The thermal resistance of the internal surface accounts for the convective and radiative heat transfer between the surface and the indoor environment. Its value can be taken as 0.13 m^{2}K/W [21]. In the presence of additional material layers on the inside of the thermal storage mass, e.g. plastering, the thermal resistance of those layers should also be included.

The indirect solar heat gains through the solar wall can be calculated, in accordance with the common equation of EN ISO 13790 [20], as:

$$Q_{gn,sw} = F_{sh,ob} A_{sol,sw} I_{sol} t / 1000. \tag{4}$$

where: $F_{sh,ob}$ – shading reduction factor for external obstacles (including terrain, nearby buildings, fixed external shading devices, etc.), [-]; I_{sol} – mean total solar irradiation on a surface with the corresponding slope and azimuth, [W/m^2]; $A_{sol,sw}$ – effective collecting area of the solar wall, [m^2].

The effective collecting area is expressed as a geometric factor, but its value depends on both geometric and non-geometric parameters. It represents the fraction of incident solar radiation that reaches the indoor environment as heat gains, after all reductions due to shading, reflection and heat losses to the outdoor environment:

$$A_{sol,sw} = F_{sh,gl} \left(1 - F_F\right) g_w \alpha_{ab} A_{sw} \eta_{sw}. \tag{5}$$

where: $F_{sh,gl}$ – shading reduction factor for movable shading provisions, [-]; F_F – frame area fraction (ratio of the projected frame area to the overall projected area of the glazing), [-]; g_w – total solar energy transmittance of the glazing unit, [-]; α_{ab} – solar absorptivity of the absorber coating, [-]; A_{sw} – internal surface area of the wall, [m^2]; η_{sw} – effective solar transmittance of the wall, [-].

The external heat loss coefficient accounts for all mechanisms of heat transfer occurring in the direction from the external surface of the wall, i.e. the absorber, to the outdoor environment. Appropriate values for the external heat loss coefficient have been extracted from the simulation outcomes using CART analysis. The most significant parameters, determining the value of this coefficient, are the infrared emissivity of the absorber coating and the type of glazing. The obtained values are given in Table 3.

The effective solar transmittance of the wall is a dimensionless factor, representing the fraction of absorbed solar radiation which reaches the indoor environment. It therefore accounts for the absorbed solar heat which is lost to the outdoor environment. Theoretically, this factor can vary between 0 and 1, while the values obtained from all simulations are in the range of 0.21 to 0.87 (with a median of 0.54). The following equation for calculation of η_{sw} was obtained through nonlinear regression analysis:

$$\eta_{sw} = \eta_{sw,0} - \left(0.105 - 0.026 k_p\right) ln\left(\tau_p\right). \tag{6}$$

$$\eta_{sw,0} = 0.915 - 0.227 ln\left(U_e\right). \tag{7}$$

Table 3: Values of the external heat loss coefficient – statistical summary.

Absorber coating	Glazing	External heat loss coefficient, U_e, [W/(m^{2}K)] Average value	Median value	Median absolute deviation
Non-selective (ε_{ab} = 0.87)	Single clear	3.2	3.3	0.30
Non-selective (ε_{ab} = 0.87)	Double clear	2.1	2.1	0.16
Selective (ε_{ab} = 0.11)	Single clear	1.1	1.1	0.08
Selective (ε_{ab} = 0.11)	Double clear	0.9	0.9	0.06

where: k_p – thermal conductivity of the wall, [W/(mK)]; U_e – external heat loss coefficient, [W/(m^{2}K)]; τ_p – time-constant of the wall, which indicates its thermal inertia, [h]:

$$\tau_p = (\rho c)_p L_p^2 / (3600 k_p). \tag{8}$$

where: L_p – thickness of the wall, [m]; $(\rho c)_p$ – volumetric heat capacity of the thermal storage mass, [J/(m^{3}K)].

For consistency, eqn (7) was designed to use the same values for the external heat loss coefficent as eqn (3), i.e. the values given in Table 3. It should be noted, however, that the actual mean value of the external heat loss coefficient will be higher in the presence of solar irradiation, due to the increased temperature of the absorber. A statistical summary of the nonlinear regression model of eqn (6) is shown in Table 4. The functional relationships of eqns (6) and (7) are illustrated graphically in Figures 4 and 5.

During the heating season, some heat gains enter the building when no heating is needed. They may lead to an undesired increase of the internal temperature above the set-point, i.e.

Table 4: Statistical summary of the effective solar transmittance model.

Fitted equation: $\eta_{sw} = \zeta_1 - \zeta_2 ln(U_e) - (\zeta_3 - \zeta_4 k_p) ln(\tau_p)$.

		95% Confidence limits				
	Value	Lower	Upper	Standard error	t-value	Pr(>|t|)
ζ_1	0.9150	0.91472	0.91532	0.00015	5996	$< 2 \times 10^{-16}$
ζ_2	0.2274	0.22716	0.22758	0.00011	2127	$< 2 \times 10^{-16}$
ζ_3	0.1046	0.10442	0.10469	0.00007	1519	$< 2 \times 10^{-16}$
ζ_4	0.0265	0.02644	0.02663	0.00005	558.4	$< 2 \times 10^{-16}$

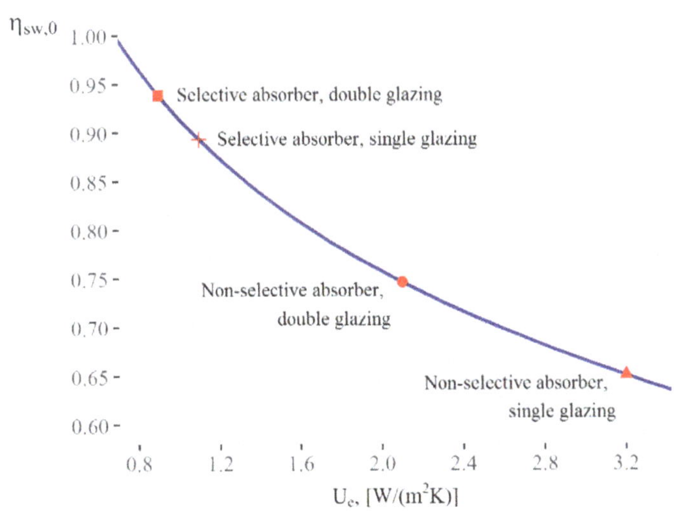

Figure 4: Variation of effective solar transmittance with external heat loss coefficient.

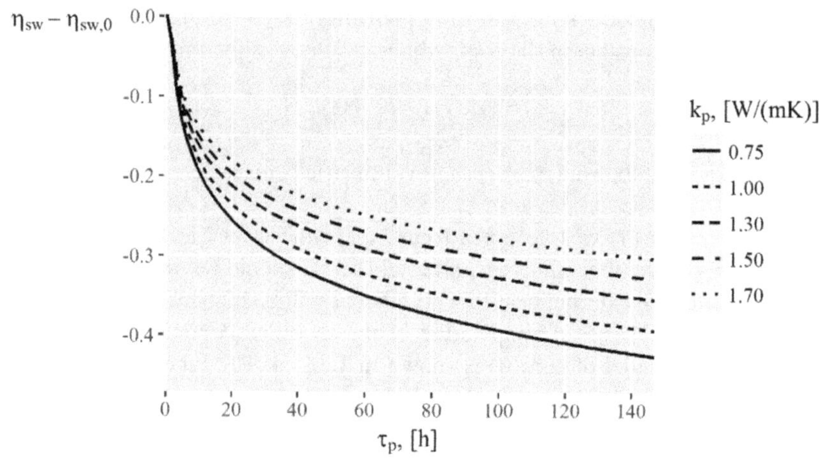

Figure 5: Variation of effective solar transmittance with thermal mass properties.

overheating. Such extra heat gains do not contribute to the reduction of building heating demand. The amount of heat gains, which are utilized for reducing the heating demand, depends on the total heat capacity of the respective thermal zone in the building. The higher the heat capacity – the greater the capability of the building to utilize the incoming heat gains (thermal storage enables a more even distribution of the heating loads in time). An analogous situation occurs in the summer when not all heat losses contribute to reduction of cooling demand. Therefore, according to the monthly method of EN ISO 13790 [20], for each building thermal zone and each calculation step (month), the energy demands for space heating, $Q_{H,nd}$, and cooling, $Q_{C,nd}$, are calculated as:

$$Q_{H,nd} = Q_{H,ht} - \eta_{H,gn} Q_{H,gn}. \tag{9}$$

$$Q_{C,nd} = Q_{C,gn} - \eta_{C,ls} Q_{C,ht}. \tag{10}$$

where: Q_{ht} – total heat transfer, [kWh]; Q_{gn} – total heat gains, [kWh]; $\eta_{H,gn}$ – dimensionless utilization factor for heat gains; $\eta_{C,ls}$ – dimensionless utilization factor for heat losses.

If the presented model is used in the context of whole-building energy analysis with the monthly method EN ISO 13790, heat losses and heat gains through the solar should be calculated separately and then each should be included in the building thermal zone's sum total of heat losses and heat gains. The heat capacity of the wall should be accounted for when calculating the internal heat capacity of the building thermal zone. The latter is used for computation of the dimensionless heat gains and heat losses utilization factors, as described in the standard [20].

Figure 6 shows comparison between the steady-state model and the test set. The plot on the left side shows the values obtained for the net heat transfer (eqn (1)) per unit area of the solar wall, with each data point representing one of all simulated solar wall configurations. In general, the observations are within an acceptable range from the diagonal line (which represents ideal coincidence). The histogram on the right side shows the distribution of the relative error. The histogram shows that for more than 90% of all data points the relative error is <15%. The median value of the relative error over the whole test set is 3.4%. However, this does not

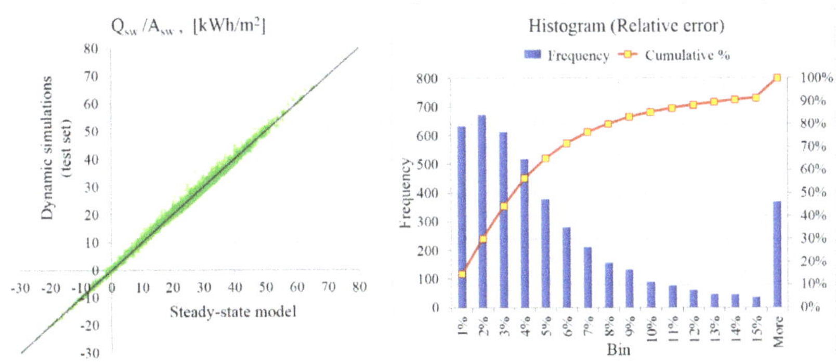

Figure 6: Accuracy of the steady-state model on the test set.

include errors which would arise when calculating the total solar energy transmittance of the glazing and the shading reduction factors of eqns (4) and (5).

The overall heat transfer coefficient and the effective solar transmittance are key factors for the energy performance of the solar wall. Both eqns (3) and (7) feature the external heat transfer coefficient, as well as the thermal conductivity, volumetric heat capacity and thickness of the thermal storage mass. The net heat transfer, as defined in eqn (1), is much more sensitive to the external heat loss coefficient than to the properties of the thermal storage material. A 10% change in the value of the thermal conductivity or the volumetric heat capacity of the wall results in only about 3% change in the output value, while a 10% change in the external heat loss coefficient results in more than 15% change in the output. The type of absorber coating and the type of glazing greatly influence the external heat loss coefficient and thereby the energy performance of the solar wall. It can be substantially improved if a selective absorber coating is used. Analysis of the experimental outcomes shows that, in such a case, the solar wall can significantly outperform a well-insulated external wall [16]. If the coating is not selective, it should be combined with double glazing. Configurations with no absorber coating and single glazing have very low efficiencies in moderate and cold climates.

On the other hand, the distance between the wall and the glazing unit exhibits little influence on the external heat loss coefficient in the range of investigated values. However, this distance should preferably be limited to 2–5 cm, to minimize shading on the absorber (i.e. reduce the value of F_F in eqn (5)). While the properties of the thermal storage material do not have a significant influence on the net heat transfer (results become slightly better when increasing the thermal resistance), they have a significant influence on the utilization of heat gains. In general, but not always, better overall performance should be obtained when the time-constant and the heat capacity of the thermal storage mass are higher [16]. The optimal time-constant may be dependent on the desired distribution of heat gains in time (a higher time-constant results in slower response to temperature variations).

For optimal performance, the solar wall should be facing south (in the Northern hemisphere) and to the maximum extend remain unshaded during the winter. Deviation of the surface azimuth of ±45° from due south reduces the efficiency of the solar wall by about 20% [16]. During the summer, however, it is necessary to provide efficient shading of the unvented solar wall to prevent undesirable heat gains. Fixed (e.g. overhangs) or movable exterior shading devices can by applied for this purpose, as well as natural shading if possible. A movable external shading device would be the most efficient option, as it is the most controllable, while an overhang would be a simpler and cheaper solution. If an overhang is used, the projection length and angle should be carefully selected, because there will be a trade-off between allowing more solar radiation to pass in the heating season and allowing less in the summer. The optimal geometry of an overhang depends on the geometry and orientation of the solar wall, as well as the geographic latitude of the site. With good shading provided, the heat capacity of the wall could (if possible) be used as a device for heat load levelling, thereby reducing the required cooling capacities.

Additional insights were obtained through the physical measurements from the test site. A season-average daily profile of the temperatures in the upper half of the test site wall is shown in Fig. 7 (θ_1 and θ_7 are the external and the internal surface temperatures accordingly; the other temperatures are measured at equal distances in between; $\theta_{1(sh)}$ is the average temperature of the shaded area of the external surface). The profile was created by calculating the average values of the measured temperatures for each hour of the day over the entire heating season (22 October 2014–22 April 2015). Analysis of the measurements indicates that, as a rough estimate, the peak temperature is shifted by 20 minutes for each centimetre of added thickness. Comparison of the temperatures in the upper and lower halves of the wall demonstrates the effect of increasing the time-constant (which is about two times higher for the upper half, primarily due to the lower thermal conductivity of the clay bricks).

As heat diffusion is faster in the lower half, the amplitudes of its internal temperature are higher, while the differences between the internal and external surface temperatures are lower. As previously mentioned, increasing the thermal inertia of the wall delays and decreases the extremums of the internal surface temperature. This reduces the risk of causing thermal discomfort to the occupants of the building, which may arise due to radiant temperature asymmetry. According to ASHRAE Standard 55, the radiant temperature asymmetry should not exceed 10 K, if it is caused by a cold wall, or 23 K, if it is caused by a warm wall [22]. A histogram of the internal temperatures of each half of the wall, showing the number of hours (frequency) during which the internal surface has experienced the given temperature (bin), is shown in Fig. 8. The histogram indicates that in moderate climatic conditions, such as those in Sofia, the solar wall should not be a cause of thermal discomfort, except in rare cases. The risk will be increased if the wall is very thin and has a low time constant.

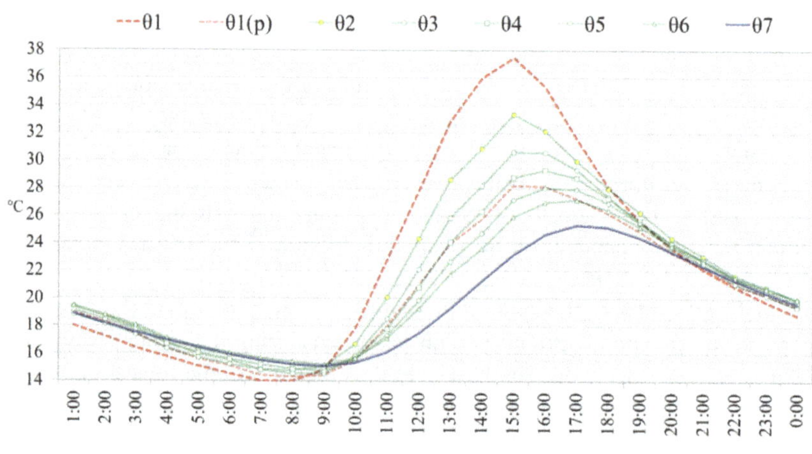

Figure 7: Season-average daily profile of the temperatures in the upper half of the wall.

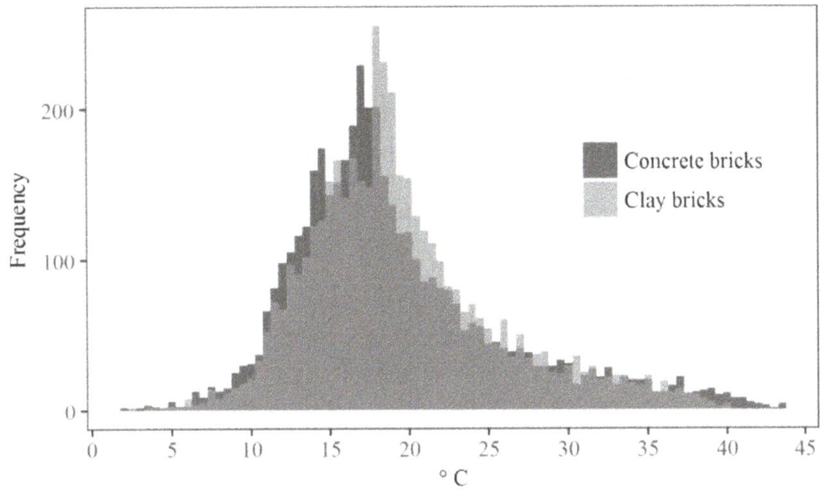

Figure 8: Histogram of the internal surface temperatures over the heating season.

An infrared picture of the internal surface of the wall is shown in Fig. 9. The picture clearly demonstrates the negative effect of the frame and dividers of the glazing unit, casting shadow on the absorber. The lower temperatures of the shaded sections of the external surface result in lower temperature regions on the internal surface (visible on the picture). This effect is amplified by the relatively high distance between the absorber and the glazing unit. As previously mentioned, this distance should preferably be limited to 2–5 cm, while frame dividers should be minimized. It is also important to reduce to the maximum extent possible the placement of furniture or other objects, which may impede the efficient heat exchange between the solar wall and the occupied space [8, 13].

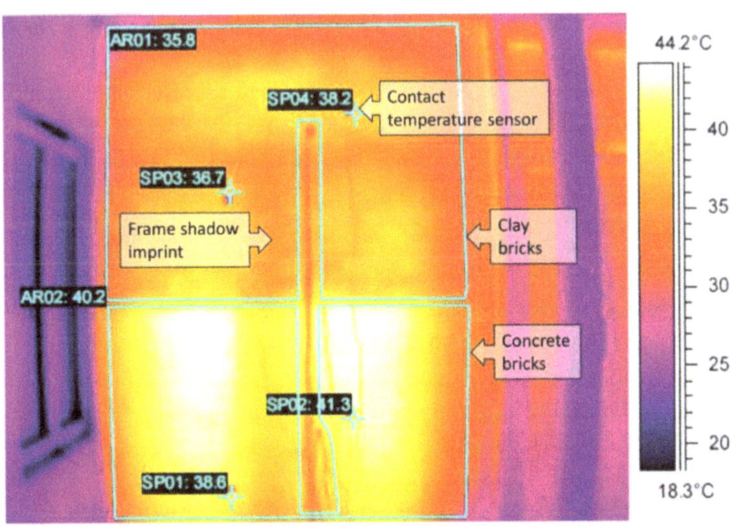

Figure 9: Infrared picture of the internal surface of the solar wall (18/02/2015 16:45).

The measured temperatures and air velocities in the air space, as well as the measured temperatures of the absorber, were used to analyse the heat transfer processes occurring between the absorber and the glazing. The analysis shows that, on the average, heat transfer due to thermal radiation was about four times more intensive than convective heat transfer in the air gap. Calculations show that the use of selective absorber (with an infrared emissivity of 0.11) would reduce the radiant heat transfer approximately seven times, which will result in about 70% reduction of total heat losses [16]. This confirms the high significance of the type of absorber coating on the energy performance of the solar wall, discussed above.

4 CONCLUSIONS

The presented simplified model can be used, along with other analysis tools, to estimate the energy and cost efficiency of an unvented solar wall, especially in the early stages of the building design process. The simplified model is robust and suitable for a wide range of climatic conditions. It provides rather conservative results and should not overestimate the efficiency of a typical solar wall. The results of the study show that in temperate and warm climates, the application of solar walls can improve building energy performance, if they are designed according to the best practices and are well combined with other elements of passive solar design.

ACKNOWLEDGEMENTS

This work has been supported by The National Science Fund of Bulgaria under projects number ДУНК-01/3 (DUNK-01/3) and ДФНИ E 02/17 (DFNI E 02/17).

REFERENCES

[1] Trombe, F. & Michel, J., *Naturally air-conditioned dwellings*, U.S. Patent 3832992, 1972.

[2] Morse, E.S., *Warming and ventilating apartments by sun's rays*, U.S. Patent 246626, 1881.

[3] Duffie, J. & Beckman, W., *Solar engineering of thermal processes*, 3rd ed, Wiley: Hoboken, New Jersey, pp. 733–759, 2006.

[4] Wilson, A., *Thermal storage wall design manual*, New Mexico Solar Energy Association: Albuquerque, New Mexico, 1979.

[5] Monsen, W.A., Klein S.A. & Beckman, W.A., The un-utilizability design method for collector-storage walls. *Solar Energy*, **29**, pp. 421–429, 1982. https://doi.org/10.1016/0038-092x(82)90079-2

[6] Saadatian, O., Sopian, K., Lim, C.H., Asim, N. & Sulaiman, M.Y., Trombe walls: A review of opportunities and challenges in research and development. *Renewable and Sustainable Energy Reviews,* **16**, pp. 6340–6351, 2012. http://dx.doi.org/10.1016/j.rser.2012.06.032

[7] De Gracia Cuesta, A., *Thermal analysis of a ventilated facade with phase change materials (PCM)*, PhD thesis, University of Lleida, Spain, 2013.

[8] Torcellini, P., Long, N., Pless, S. & Judkoff, R., *Evaluation of the low-energy design and energy performance of the Zion National Park Visitors Center,* NREL/TP-550-34607, NREL: Golden, Colorado, 2005.

[9] Torcellini, P. & Pless, S., Trombe walls in low-energy buildings: Practical experiences, Presented at *8th World Renewable Energy Congress and Expo*, Denver, Colorado, 2004.

[10] Judkoff, R. & Sokol, F., Performance of a selective-surface Trombe wall in a small commercial building, Presented at *AS/ISES Annual Meeting*, Philadelphia, Pennsylvania, 1981.

[11] Steven Winter Associates, *The passive solar design and construction handbook*, Wiley: New York, 1998.

[12] Dowson, M., Harrison, D. & Dehouche, Z., Trombe walls with nanoporous aerogel insulation applied to UK housing refurbishments. *International Journal of Smart and Nano Materials*, **5**(4), pp. 283–303, 2014. http://dx.doi.org/10.1080/19475411.2014.999730

[13] Torcellini, P., Pless, S., Judkoff, R. & Crawley, D., Solar technologies & the building envelope. *ASHRAE Journal*, **49**(4), pp. 14–22, 2007.

[14] Klein, S.A., Beckman, W.A., Mitchell, J.W., Duffie, J.A., Duffie, N.A., Freeman, T.L., et al., *TRNSYS 17: A transient system simulation program*, University of Wisconsin-Madison, 2010.

[15] TESS Component Library Package. Available at: http://www.trnsys.com/tess-libraries

[16] Stankov, B., *Models of the heat transfer processes in passive solar systems*, PhD thesis, Technical University of Sofia, Bulgaria, 2015.

[17] *Subroutine Type36*, Solar Energy Laboratory, University of Wisconsin-Madison, 2013.

[18] ANSI/ASHRAE/IESNA Standard 90.1-2007, *Energy standard for buildings except low-rise residential buildings*, Atlanta, Georgia, 2007.

[19] Kottek, M., Grieser, J., Beck, C., Rudolf, B. & Rubel F., World map of the Köppen-Geiger climate classification updated. *Meteorologische Zeitschrift*, **15**, pp. 259–263, 2006. https://doi.org/10.1127/0941-2948/2006/0130

[20] EN ISO 13790:2008, *Energy performance of buildings -- Calculation of energy use for space heating and cooling*, Geneva, Switzerland, 2008.

[21] *CIBSE Guide A – Environmental design*, 8th ed., The Chartered Institution of Building Services Engineers: London, United Kingdom, 2015.

[22] ANSI/ASHRAE Standard 55-2010, *Thermal environmental conditions for human occupancy*, Atlanta, Georgia, 2010.

EFFICIENT ENERGY MANAGEMENT SYSTEM AT KUWAIT OIL COMPANY, KUWAIT, A CASE STUDY

YOUSEF E. AL-QALLAF, FATIMA OWAYED & PARISUDDA RAO
Health & Environment Team, KOC, Kuwait.

ABSTRACT

Effective energy management involves making decisions that lead to the conservation of energy and the efficient use of resources for sustainable future. Kuwait Oil Company (KOC), a subsidiary of Kuwait Petroleum Corporation (KPC), is involved in exploration, drilling and production of oil and gas. KOC is fully committed towards energy management, energy efficiency and greenhouse gases (GHGs) emissions reduction, which may help in minimizing energy costs and mitigating environmental effects. In order to meet national and international standards ISO 50001 Energy Management System (EnMS), KOC undertaken a pilot study for developing an effective energy management program for KOC representative process units and main buildings. The objective of the program was to create an energy baseline and identify the potential improvement areas and provide inputs for the implementation of ISO 50001 for certification. KOC has established the Energy Performance Indicators (EnPIs) for each of the process units and specific KPIs has been identified to monitor and control the energy performance. Furthermore, the study highlights the major achievements towards energy management, energy efficiency and greenhouse gas (GHGs) emissions reduction in order to help in minimizing energy costs and mitigating environmental effects.

Key Words: energy baseline, energy cost & energy saving best practices, EnPIs, ISO EnMS.

1 INTRODUCTION

The systemic waste of natural resources in the Gulf is eroding economic resilience to shocks and increasing security risks. The six Gulf Cooperation Council (GCC) countries now consume more primary energy than the whole of Africa. Almost 100% of energy is produced from oil and gas without carbon dioxide abatement. If the region's fuel demand were to continue rising as it has over the last decade, it would double by 2024.

This is a deeply undesirable prospect for both the national security of each state and the global environment. KOC was established in 1934, is one of the largest oil exporter in the world, headquartered in Ahmadi, Kuwait. The Company activities had extended to include exploration operations, on-shore and offshore surveys, drilling of test wells, and developing of producing fields in addition to crude and natural gas exploration.

KOC undertaken a pilot study consisting of three process units and two main buildings. The task was to create an energy baseline & identify the potential energy saving opportunities.

All energy sources have been considered in the study and wherever the data was available have been analyzed for performance and the opportunities in the five selected pilot sites. They have conducted a detailed study after site visits and interactions with the process owners and identified the potential areas for improvement with recommendation. The selected pilot sites are mentioned below [1].

- Gathering Center (GC-04)
- Booster Station (BS-160)
- Effluent Water Disposable Plant (EWDP – II)
- KOC Main Office Building
- KOC New Office Complex Building (B1)

DOI: 10.2495/EQ-V2-N4-352-359

The program was developed in accordance with national and internal standards ISO 50001:2011 Energy Management System (EnMS) Requirements [2]. KOC has already appointed a Management Representative, Core energy management team to guide and take forward the energy improvement activities. Energy Policy has been defined. Training on ISO 50001 awareness has been provided to the employees & contractors.

2 METHODOLOGY

3 COMPLIANCE TO LEGAL & OTHER REQUIREMENTS

The study was comply with national and international regulations on energy conservation code for buildings [3]. Further, the State of Kuwait State of Kuwait is a signatory to COP#21 wherein it has committed to reduce the impact on environment by reducing the greenhouse gases [4]. To achieve this KOC has to reduce its burning of fossil fuel.

4 BASELINE

Based on the data made available, the baseline for all the five pilot sites have been made and analyzed. The data period has been considered from Oct'2013 to Mar'2015 (18 months period). The Target for each site has been derived using the six-sigma concept considering a confidence level of 70%. Energy Conservation Plan that has been identified [5] is addressed in the Energy savings and cost effective opportunities.

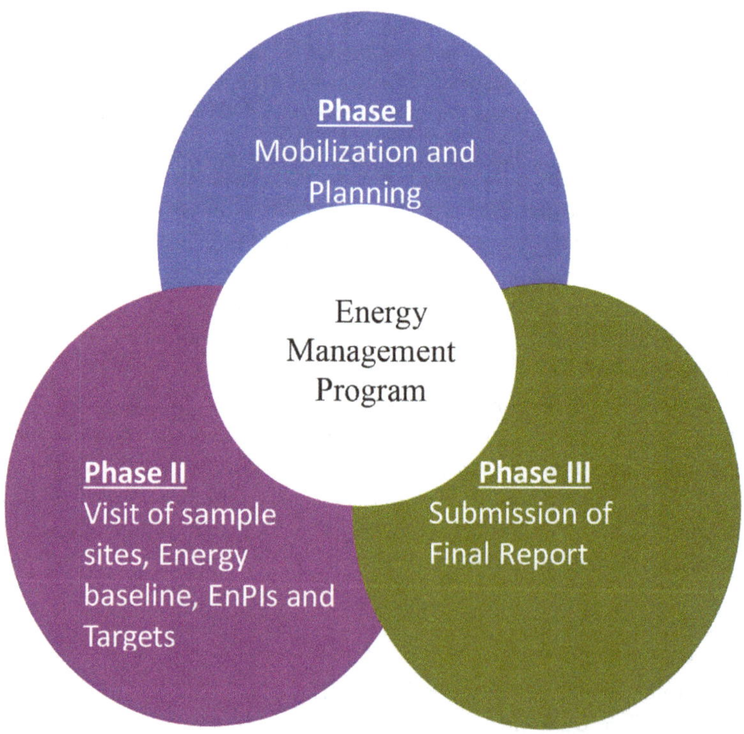

Figure 1: Different Phases for Energy Management Program.

Table 1: Baseline Data Including Energy Performance Indicators (EnPIs)

S.No	Pilot Site	Baseline (Oct'13 - Mar'15)		EnPI	Estimated Energy Savings %
1	GC 04	Electrical	613.5	kWh/MBOE	**15.2**
		Thermal	12934.5	SCF/MBOE	**7.0**
		Overall energy Equivalent	3684	MkCal/MBOE	**7.7**
2	BS 160	Electrical	1118.7	kWh/MBOE	**20.7**
		Thermal	4037.4	MSCF/MBOE	**3.5**
		Overall energy Equivalent	986.1	MMkCal/MBOE	**3.5**
3	EWDP II	Electrical	0.413	kWh/BBLS	**5.7**
		Thermal	0.501	SCF/BBLS	**10.8**
		Overall energy Equivalent	0.477	MkCal/BBLS	**4.2**
4	Main Office Building	Winter	2.47	kWh/m^2/day	**4.98**
		Summer	2.90	kWh/m^2/day	**6.25**
5	New Office Complex	Winter	0.39	kWh/m^2/day	**11.1**
		Summer	1.09	kWh/m^2/day	**12.3**

4.1 Identification of areas of Significant Energy Use (GC-04)

Significant energy use refers to energy use accounting for substantial energy consumption and/or offering considerable potential for energy performance improvement. In line with this, significant energy use equipment/processes were identified based on the electrical load data furnished for GC-04. The share of total connected load of equipment & machineries, which are presently operated are depicted below to understand and highlight the significant power consumers [6].

4.2 Identification of areas of Significant Energy Use (BS-160)

Significant energy use refers to energy use accounting for substantial energy consumption and/or offering considerable potential for energy performance improvement. In line with this, significant energy use equipment/processes were identified based on the electrical load data furnished for BS – 160 [7]. The share of total energy use is evolved to understand the contribution of thermal energy (fuel gas) and electrical energy in the day-to-day operations of BS 160 is depicted in figure below.

4.3 Identification of areas of Significant Energy Use (EWDP-II)

The thermal energy is used only in sludge separators and rest of other systems utilizes electrical energy and majority of them are motor driven equipment. The share of total connected

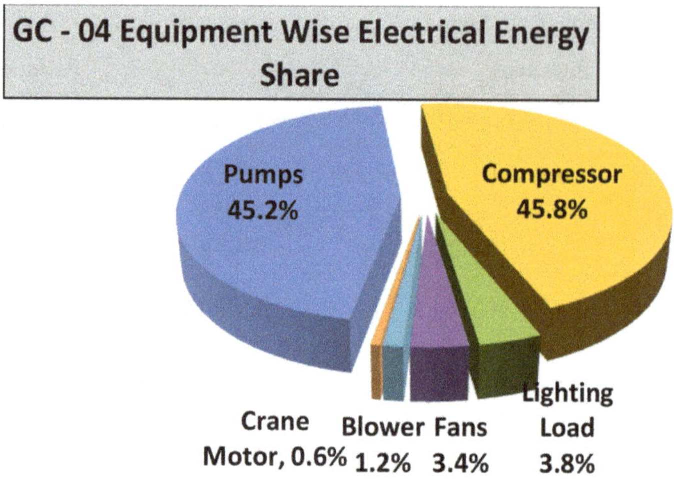

Figure 2: Significant Energy Consumption for GC-04.

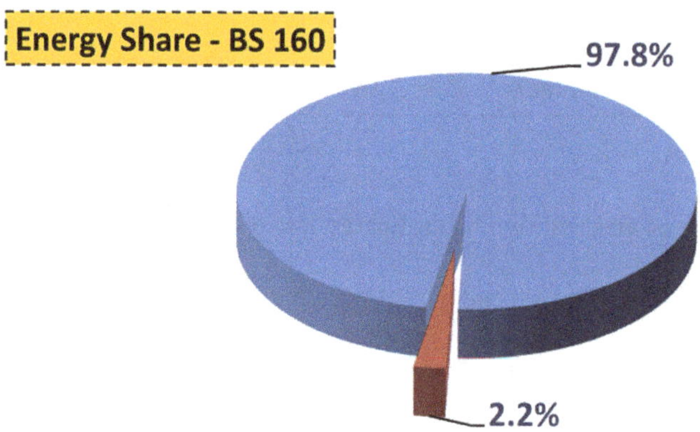

Figure 3: Significant Energy Consumption for BS-160.

load of equipment & machineries, which are presently operated are depicted to understand and highlight the significant power guzzlers.

4.4 Identification of areas of Significant Energy Use (Main Office Building)

Significant energy use equipment/processes were identified based on the electrical load data furnished for main office building. The share of total connected load of electrical equipment & machineries.

4.5 Identification of areas of Significant Energy Use (New Office Complex)

The major variables affecting the energy use for new office complex has identified based on observations, analysis and understanding of process operations and energy utilization [8].

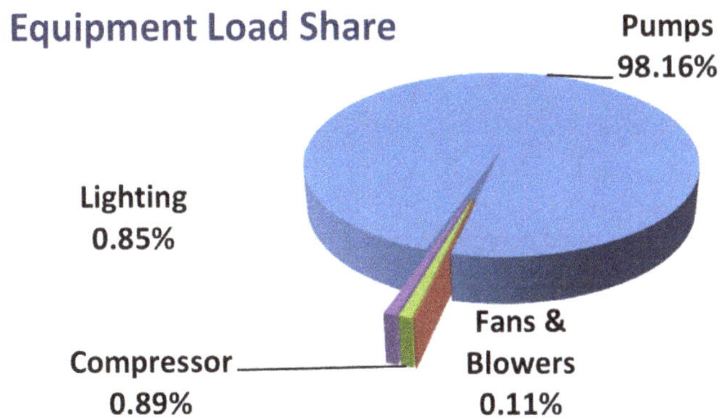

Figure 4: Significant Energy Consumption for EWDP-II.

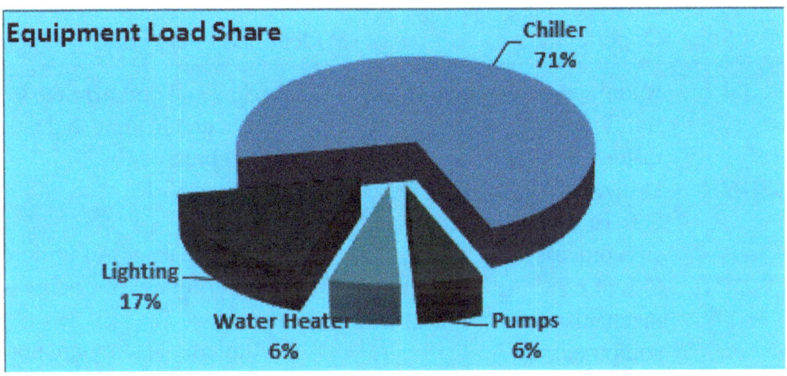

Figure 5: Significant Energy Consumption for Main Office Building.

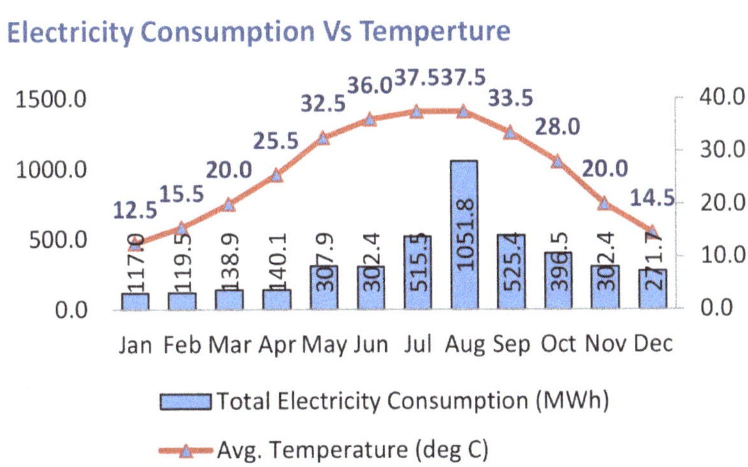

Figure 6: Significant Energy Consumption Vs Temperature at New Office Complex.

Further the trend for electricity consumption vs ambient temperature is evolved for the baseline period to visualize the impact of climatic variations on electricity as depicted in the below [9].

5 ENERGY SAVING OPPORTUNITIES

The site-specific observations comprises identification of significant energy use discussed above and existing & identified energy savings opportunities are summarized for all five pilot sites in below table [10].

6 CONCLUSION

The company should creating baseline for other process units, in order to monitor and verify the energy performance. Further, metering & sub metering system for electrical and thermal where

Table 2: Existing and Identified Energy Saving Opportunities

S.No	Pilot Site	Existing Energy Savings Opportunities	Identified Energy Saving Opportunities
1	GC 04	Running Gas engine/turbine for Tank Vapor compressors instead of motors	1. Review and optimize the sizing of Pumps and loading and unloading of compressors.
2	BS 160	Utilization of Energy Efficient Gas Turbine in gas compression process	2. Installing VFD based on Process demands. 3. Rectify Leakages of utilities.
3	EWDP II	Use of metering & monitoring mechanisms for the electrical equipment	4. Monitoring use of Energy with Meters 5. Optimizing loading on motors. 6. Monitor Carbon and water foot print in all process areas 7. Optimize transformer losses 8. Use LED instead of conventional lamps 9. HVAC usage and control
4	Main Office Building	1. LED Replacement 2. Switching off office loads (computers, lights, AC units etc.) When office is not occupied. 3. VFD pump sets are installed to modulate chilled water flow. 4. Water saving devices are installed on all taps.	1. Insulate all cold lines and vessels 2. Optimize Insulation thickness and material 3. Optomize air volumes with false ceiling and segregation of areas 4. Usage of Air curtains 5. Minimize roof heating loads with roof cooling and roof painting 6. Precooling of fresh air with air to air heat exchangers 7. Optimize thermostat settings 8. Regular maintenance of A/C and Refrigeration 9. Minimize part load operations
5	New Office Complex		

not available, be provided on priority. (Energy meters and Flow meters). A comprehensive investment grade energy audit (IGA) should be conducted to for each process site and buildings for capturing the individual equipment performance and identify the losses and pin point energy conservation opportunities and also Carry out Testing, Adjusting and Balancing (TAB) for the HVAC system for the buildings to improve the chiller efficiency and performance.

Additionally, increase the use of renewable and clean energy technologies such as Solar PV in office buildings (use of Solar PV for lighting load in main & new office buildings to replace the conventional power source), Solar Thermal technologies in process heating and use of LNG instead of fossil fuel based electricity. For the Green building concept, KOC needs to get certified to ISO 50001 and then follow the LEED reference guide for Building Design & Construction v4 [11].

ABBREVATIONS

KOC – Kuwait Oil Company
KPC – Kuwait Petroleum Corporation
GHGs – Greenhouse Gases
EnMS – Energy Management System
COP – Conference of Parties
GC- Gathering Center
BS – Booster Station
EWDP – Effluent Water Disposal Plant
EnPI – Energy Performance Indicator
GCC - Gulf Cooperation Council
LED – Light-Emitting Diode
VFD – Variable Frequency Drive
IGA - Investment Grade Energy Audit
HVAC – Heating Ventilation Air Conditioning System
TAB - Testing, Adjusting and Balancing
LNG – Liquefied Natural Gas
LEED - Leadership in Energy and Environmental Design

REFERENCES

[1] A Study on Energy Management Program in KOC.
[2] ISO 50001:2011 Energy Management System (EnMS).
[3] Related Standards and Guidelines on Energy Conservation Code in Buildings by (Ministry of Energy (MoE), Kuwait
[4] State of Kuwait State of Kuwait is a signatory to COP#21 wherein it has committed to reduce the impact on environment by reducing the GHG emissions.
[5] Energy Conservation & Demand Management Plan 2014 – 2019 Technical Report – OAKVILLE.
[6] Energy Efficiency Reference for Environmental Reviewers, March 2010 – EPA Publication No. 315-R-09-001.
[7] Bhattacharyya, S.C., Energy data and energy balance. *Energy Economics*, Springer-Verlag, London, 2011.
https://doi.org/10.1007/978-0-85729-268-1_2

[8] Kaneda, D., Jacobson, B. & Rumsey, P., *Plug oad Reduction: The Next Big Hurdle for Net Zero Energy Building Design*. ACEEE Summer Study on Energy Efficiency in Buildings; 2010.

[9] Menezes, A., Cripps, A., Bouchlaghem, D. & Buswell, R., *Analysis of electricity consumption for lighting and small power in office buildings*, CIBSE Technical Symposium, 6th–7th September, DeMontfort University, Leicester, UK, 2011.

[10] IPIECA, Climate Change 2013 – "Saving energy in the oil and gas industry", 2013. Available at: www.ipieca.org

[11] Available at: http://www.usgbc.org/leed

SUSTAINABLE MANAGEMENT MODEL BASED ON RENEWABLE ENERGIES FOR THE FIRST CAPITAL OF THE CALIFORNIAS, LORETO, MEXICO

PEDRO-DANIEL MOLINA-PFENNIG & OSCAR ARIZPE COVARRUBIAS
University of Baja California Sur, Mexico.

ABSTRACT

The Mexican state of Baja California Sur has a high rate of population growth. It is also one of the states that are most vulnerable to climate change. Due to its location on the southern side of a roughly 900-mile long peninsula, and its natural separation from mainland Mexico, its power transmission networks are completely independent of the rest of the country. Thus, nearly all the energy used to generate electricity must be shipped to the state in the form of fossil fuels. The importation of energy supplies from the mainland results in higher costs for the state than in other areas of the country, causes greater environmental damage, and prevents a steady supply of energy to the state. This study's objective is to propose a sustainable management model and to provide a reference to feasible sites available that could serve the Loreto region. An analytical model has been developed with multiple criteria and geographic information systems. This will allow for a wide range of spatial analysis of information covering the calculation of slopes, orientation, irradiation, infrastructure, etc. The municipal region of Loreto has roughly 288 square kilometres of land deemed suitable for the installation of solar plants. This area comprises 1.62% of the municipality. In 2016, the maximum electrical power demand for the entire state of Baja California Sur was 628 Megawatts per hour according to the Federal Electricity Commission (CFE). Loreto's electrical capacity is currently 17 MWh. Based on calculations that one photovoltaic plant located on two acres of land can produce one MWh, solar plants in the region could, theoretically, produce up to 14,403.35 MWh. Clearly, this potential capacity would be well above the demands of the municipality, which encompasses 3.8% of the state territory.

Keywords: *Analysis, Capital of The Californias, Gis, Loreto, Multi-Criteria, Solar Radiation*

1 INTRODUCTION

In its Executive Summary of 2015, the International Energy Agency highlights the significant increase in the generation of renewable energy from non-fossil sources in the world. Renewable energy currently makes up 19% of the total share of generated global energy. This percentage reflects the large investments made in renewables in 2014, of 270,000 million dollars (270 billion USD). It is estimated that by 2030, global demand for energy will increase by 34%. However, most of the world's energy production – 80% – is still expected to come from fossil fuels [1]. As a result, significant – and dangerous – levels of carbon emissions will continue to be produced in the years ahead. To reduce emissions, a rapid development of renewable energy production around the world is underway and reflects profound strategic changes in the way energy systems are now being produced. These important changes increasingly focus on creating diverse, efficient, and sustainable energy systems. In addition, the commitments made at the Kyoto Summit, to mitigate climate change through the reduction of carbon emissions, have yet to be fully implemented.

Mexico's transition from traditional energy sources to renewables will provide great opportunities for investors to tap into the country's enormous potential for developing solar, wind, geothermal, hydroelectric, bioenergy, tidal, and other energy sources. In Baja California Sur, the average rates of solar radiation range between 4.4 kwh/m^{2}day and 6.9 kwh/m^{2}day [2]. In contrast, the Loreto region produces an average of 5.2 kwh/m^{2}day [3], which makes it

DOI: 10.2495/EQ-V2-N4-360-369

Table 1: Methodological steps.

1.	Identification of variables
2.	Classification by levels of suitability
3.	Assignment of value to variables
4.	Generation of digital maps of the variables in Quantum GIS software
5.	Intersection of values, overlapping of layers and calculation of suitable areas
6.	Simulation of the model

particularly attractive for establishing photovoltaic energy production. Moreover, Loreto's location and high solar energy potential are important factors for the state when it considers policy measures and actions to achieve sustainability, as well as decisions regarding the exploitation of its region's vast energy sources.

2 METHODOLOGY

The methodological process of this study uses a conceptual approach that follows the diagram flow (below) in its six stages. The first three stages are identified as areas of importance, list various conditions used to determine technical, economic, legal, and social feasibility, and are based on specific criteria. These stages can be extended or modified, depending on the needs and conditions required in a chosen study area.

A digital elevation model with a 15-m resolution used by INEGI, Mexico's national geographical and statistics agency is employed and features topographical maps of an area's level of suitability in Loreto, referred to as UTM zone 12. The maps are scaled at 1:50,000 and are in an SHP form. These maps were cut and selected according to variables listed above and show the municipality as one complete unit, since some maps will intersect with other municipal maps. Transitioning from raster layers to vector layers of orientation area's and slopes facilitates the superposition of other vector layers. To determine the area's levels of suitability, maps can be superposed and will indicate their relative values. The layers that indicate protected and non-viable areas serve as boundary markers that must be taken into account when considering Mexico's environmental protection laws. The illustrations indicate where the most relevant aspects of the selected variables coincide and show areas most suitable for photovoltaic installations, according to criteria established in the following tables.

By using an impact/suitability model, formulated by Gómez Orea [4], an area's potential photovoltaic capacity can be analysed. Suitability is defined as the set of conditions required for a location and the variability of factors that produce an impact on new activities. This is a fundamental part of the procedure and is based on analysis and spatial modelling from the GIS proposed by Fischer and Nijkamp [5], and the planning tools of Birken [6] and Lavalle [7]. Feedback obtained from the dynamic model of Eastman *et al.* [8] will also improve assessment techniques.

3 RESEARCH RESULTS

3.1 Identification of variables

Table 2: Identification of variables.

Category	Variables	Key
Infrastructure	Communication routes	Vs
	Electrical transmission lines	Rt
Area	Slope	Pe
	Orientation	Or
Climate	Solar irradiation	Irs
	Temperature	Tp
Conservation restrictions	Protected areas	Ap
	Hydrological protection zone	Ph
	Archaeological sites	Sa
Management constraints	Type of property	Tp
	Land use and vegetation	Usv
	Flood prone zones	AI
Natural phenomenon restrictions	Hurricanes	Hu
	Earthquakes	Si

3.2 Classification by levels of suitability

Table 3: Classification by levels of suitability

Suitability levels	Features
Level (5) very high	The characteristics of the variables meet the criteria
Level (4) high	The variables almost entirely meet the criteria
Level (3) middle	Midpoint between the areas that meet the criteria
Level (2) low	The conditions are limited
Level (1) none	The conditions are not considered appropriate

3.3 Assignment of values to variables.

Electrical transmission lines: In order to protect solar energy structures, electrical lines must be at least 50 m from solar plant installations. These installations must also be located within a 10 km (6.2 miles) radius of electrical transmission lines (Dominguez, 2002) [9].

Table 4: Suitability of network transmission

Infrastructure	
Network transmission (in meters)	
Suitability levels	**Features**
Level (5) very high	150 m to 1000 m
Level (4) high	From 1000 m to 3000 m
Level (3) middle	From 3000 m to 7000 m
Level (2) low	7000 m to 10000 m
Level (1) none	More than 10000 m

Road network: Two important constraints must be considered: One must focus on safeguarding the solar installation structure and the other on protecting nearby structures or areas. These structures must not be located within 150 m of a roadway or at a distance greater than 5 km (3.1 miles) (Rodriguez, 2010) [10].

Table 5: Suitability of road network.

Infrastructure	
Road distances	
Suitability levels	**Features**
Level (5) very high	150 m to 1000 m
Level (4) high	1000 m to 2000 m
Level (3) middle	From 3000 m to 4000 m
Level (2) low	5000 m to 4000 m
Level (1) none	More than 5000 m

Slopes: Although solar structures must not be installed on slopes greater than 4°, according to Dominguez (2002) [9], in order to protect landscapes, solar installation manufacturers claim to have rotating structures that can face the sun at all times. This feature, they say, allows installation on slopes up to 20°.

Table 6: Suitability of slopes.

Area	
Slopes (in degrees)	
Suitability levels	**Features**
Level (5) very high	Slopes between 0^0 and 4^0
Level (4) high	Slopes between 10^0 and 5^0
Level (3) middle	Slopes between 15^0 and 11^0
Level (2) low	Slopes between 16^0 and 20^0
Level (1) none	Slopes that exceed 20^0

Orientation: In order to provide an effective map, an illustration of viable areas facing south is required, since it is known that south-facing land areas receive more hours of daily sunshine, and installations in these areas will generate more efficient energy production.

Table 7: Suitability of orientation

Area	
Orientation (in degrees)	
Suitability levels	**Features**
Level (5) very high	Facing South, 180^0
Level (4) high	Facing South, 135^0 between 225^0
Level (1) none	Invalid orientation

The ***solar resource*** is determined by the amount of sunshine, or insolation, which can be harnessed and converted into electricity. This is the amount of solar radiation received at a given time on a given surface area. Establishing the site of the project in an area with an excellent solar resource provides a maximum and efficient use of the soil and minimizes its environmental impact.

Table 8: Suitability of solar radiation

Climate	
Solar radiation (in kilowatt hours per square meter)	
Suitability levels	**Features**
Level (5) very high	Greater than 6.5 kWh/m^2
Level (4) high	Between 4.5 kWh/m^2 and 6.5 KWh/m^2
Level (3) middle	Between 2.5 kWh/m^2 and 4.5 kWh/m^2
Level (2) low	Between 1.8 kWh/m^2 and 2.5 kWh/m^2
Level (1) none	Less than 1.8 kWh/m^2

Average ambient temperature: Silicon cells deliver their highest voltage at 25°C. With each additional degree, performance decreases between 0.3% and 0.7%.

Table 9: Suitability of temperature

Climate	
Temperature (in Celsius centigrade)	
Suitability levels	**Features**
Level (5) very high	From 20°C to 25°C
Level (4) high	26°C/29°C
Level (3) middle	30°C to 35°C
Level (2) low	36°C to 40°C
Level (1) none	More than 40°C and less than 15°C

Lithology: This will take into account specific points of interest in the physical characteristics or formations of the area as well as considerations for natural, protected spaces and land uses, such as residential, hydrological, etc.

Table 10: Suitability of protected areas

Conservation restrictions	
Protected areas	
Suitability levels	**Features**
Level (5) very high	Areas without protection
Level (3) middle	Human settlements
Level (3) middle	Sustainable use of natural resources
Level (3) middle	Recovery
Level (1) none	Protected natural areas
Hydrological protection	
Level (5) very high	Areas without Hydrological protection
Level (1) none	Areas with Hydrological protection
Archaeological areas	
Level (5) very high	Areas that do not have archaeological sites
Level (1) none	Areas that have archaeological sites

Land uses: The lands considered as non-viable areas that must be taken into account are the following: ponds (100 m), shorelines (200 m), green spaces (100 m), beaches, dunes and sandbanks (200 m), dams (100 m), residential areas, scattered and continuous roads (500 m), and saline areas (200 m). Lands with some anthropogenic derivation have at least 100 m of protection. Areas that merit special protection status include land from a natural origin as well as land from anthropogenic sources and shall have 200 m of protection (Quijano, 2011) [11]. 500 m of protection is suggested for population centres (Rodriguez, 2010) [10].

Table 11: Suitability of land constraints

Management constraints	
Type of property	
Suitability levels	**Features**
Level (5) very high	Private individuals
Level (1) none	Land in common use
Land use and vegetation	
Level (5) very high	Unused agricultural land
Level (4) high	Marginal areas
Level (4) high	Treated land
Level (3) middle	Areas without vegetation
Level (3) middle	Halophile and induced grassland
Level (1) none	Crops, Livestock, forestry
Subject to flood land	
Level (5) very high	Flood safe zones
Level (1) none	Flood prone zones

Natural phenomena: The occurrence of hurricanes and earthquakes in the region is an important consideration when planning the construction of photovoltaic energy structures. These structures should support a minimum wind speed of 150 kmh and have a solid foundation for earthquakes.

Table 12: Hurricane and earthquake levels

Restrictions by natural phenomena	
Hurricanes (on the Saffir-Simpson scale)	
Suitability levels	**Features**
Level (5) very high	Category 0 to 1
Level (4) high	Category 1–2
Level (3) Media	Category 2–3
Level (2) low	Category 3–4
Level (1) none	Category 5
Earthquakes (in degrees Richter scale)	
Level (5) very high	Size 0–2
Level (4) high	Magnitude 2–3
Level (3) Media	Magnitude 3–4
Level (2) low	Magnitude 4 to 5
Level (1) none	Greater than 5 magnitude

4 FACTOR ASSESSMENT

4.1 Suitability considerations

This study places great importance on the selection of the most viable sites available. In fact, site selection is considered the most important factor. Other important considerations include electric transmission lines, the location of roads, slopes, orientation, and irradiation. A comparative values procedure was drawn up by Saaty (12). This is a procedure that employs a 'method of analytical hierarchy' and is used in the Expert Choice application. By applying the Super Decisions program, an assessment of the relative importance of the criteria is made with the use of the model below. A good result (an inconsistency of 0.07828), which ensures reliability, was obtained from the procedure. The inconsistency index must always be less than 0.1.

4.1.1 Formulation of the algorithm

Rt = electrical transmission lines (0.132)
Vc = roads (0.12)
Pe = slopes (0.106)
Or = orientation (0.106)
Irs = irradiation (0.1)
Te = temperature (0.099)
Ap = protected Areas (0.043)
Ph = hydrological protection (0.091)
Sa = archaeological sites (0.041)
Tp = type of property (0.056)
Usv = use of soil and vegetation (0.041)
Ai = flood prone zones (0.041)
Hu = hurricane (0.008)
Si = earthquakes (0.008)

5 EQUATION

The equation results from the sum of the weighted and normalized variables and is multiplied by the degree of suitability.

$$= \sum ((Rt(G.)+Vc(G.)+Pe(G.)+Or(G.)+Irs(G.)+Te(G.)+Ap(G.)+ Ph(G.+Sa(G.)+Tp(G.)+ Usv(G.)+Ai(G.)+Hu(G).+Si(G.)). \quad (1)$$

The main selection of the variables is first made by using an economic feasibility assessment. Then, techniques, restrictions, and environmental impacts are considered.

Weighting is required to make decisions on obtaining a value or indicators, which are classified according to their relative importance. (Romero, 1996) [13]. Standard units tend to be very different, therefore, they should be normalized according to the following formula:

Normalization of the criterion – value of the criterion/ (best-worst value)

6 SIMULATION

In order to test and adjust the model, a simulation was developed using a Quantum GIS 2.16 program. By combining a superposition of layers with previously defined variables, a functional and realistic map is created, which illustrates appropriate and feasible locations for solar plant installations.

Maps are used with various polygons in the program and reflect data drawn from different variables. A blue polygon is used to represent a buffer zone for transmission networks and roads. In this case, a 1 km buffer zone is used. Likewise, areas most suitable for slopes of 0 to 4 degrees (indicated by pink polygons), as well as south-facing areas positioned at 180° (green polygons), are also defined. Loreto is a very small region, so information regarding municipal temperatures and irradiation do not appear due to the fact that there are no significant differences in this area. When the polygons are layered over each other or intersected,

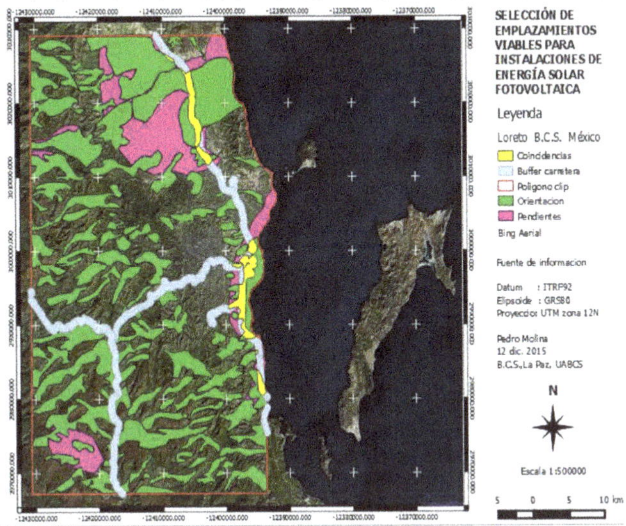

Figure 1: Partial map of feasible areas for photovoltaic energy installations in Loreto. (Source: own elaboration)

the resulting surface area (yellow polygons) emerges. The size of the combined areas suitable for solar plant installations is calculated to be 288.1 square kilometres (179 square miles, or 1.62% of the Loreto region).

There is evidence that solar field installations on two hectares (4.9 acres) can produce one megawatt per hour. Thus, it is calculated that 14,403.35 MWh could potentially be generated in this region. This amount would be well above Loreto's current electrical demand. The region's enormous potential for developing renewable energy should attract the attention of Mexico's political leaders interested in achieving environmental sustainability in the state, as well as investors eager to take advantage of these vast, untapped resources.

DISCUSSION:

Loreto does not have its own power plant, so all of its energy must be supplied by power plants in the La Paz area, 217 miles away. Having a small solar energy plant in the region would eliminate the problematic dependency on energy supplies from outside of the area. It would also provide important energy security and would significantly reduce pollution in the state. And given that the economy is based mainly on tourism, it would provide real opportunities for environmental sustainability in the region.

This model is applicable to any region, and can be modified or changed to accommodate the particular conditions of an area, as long as the following conditions to ensure a sustainable energy system are met:

1. That it is theoretically sustainable; that its construction is possible within the limits of current technology, and where the critical elements of its operational system are not compromised.
2. That it is economically viable. The project will consider environmental and social considerations as a part of its economic viability.

3. That it is socially sustainable. The community must accept the project and be willing to collaborate in its development, with the idea in mind that community participation fosters social development.
4. That it is environmentally sustainable. The renewable energy project will provide environmental benefits to the community, unlike traditional energy plants, which cause environmental damage.

REFERENCES

[1] International Energy Agency IEA 2015.

[2] Maps of Global solar irradiation in Mexico (R. Almanza E. Cajigal, S. r. j. Barrientos A. 1997) insolation of Mexico reports. Southwest Technology Development Institute, NMSU, 1999.

[3] Alberto-Jesus Perea-Moreno 1, and Quetzalcoatl Hernandez-Escobedo 2, 1 Department of Applied Physics, University of Cordoba, CEIA3, Campus of Rabanales, 14071 Córdoba, Spain 2 Faculty of Engineering Campus Coatzacoalcos, University of Veracruz, Harrapa, 0052229 Veracruz, Mexico; Published: 3 November 2016.

[4] Gómez Orea, Domingo. 1992. Planificación rural. Madrid: Agrícola Española.

[5] Fischer, M.M. & Nijkamp, P., geographic information systems and spatial analysis. *The Annals of Regional Science*, **26**(1), pp. 3–17, 1992. https://doi.org/10.1007/bf01581477

[6] Birkin, M., Elements of a Model Based Geographic Information Systems for the Evaluation of Urban Policy, 1990.

[7] LaValle, S.M. & Konkimalla, P., Algorithms for computing numerical optimal feedback motion strategies. *The International Journal of Robotics Research*, **20**(9), pp. 729–752, 2001. https://doi.org/10.1177/02783640122067633

[8] Eastman, R.J., *Guide to GIS and Image processing* Vol.2. Clark university, USA, 2001.

[9] Rodríguez, M., La Ordenación y la Planificación de las Fuentes Renovables de Energía en la Isla de Cuba desde una perspectiva territorial. Estudio de caso en el municipio de Guama a partir de un Geo portal [Libro]. - Sevilla: Doctorado Estrategias y Planificación Territorial-Facultad de Humanidades. Universidad Pablo Olavide, 2010.

[10] Domínguez, J., Los Sistemas de Información Geográfica en la Planificación e Integración de Energías Renovables [Libro]. - Madrid: CIEMAT, 2002. - Vol. I.

[11] Quijano, R., Plataforma Integrada de Modelización para la Energía, Ambiente y Desarrollo Sostenible– MODERGIS "Aplicación Caso Piloto Colombia" [Libro]. - Bogotá: Doctorado en Ingeniería-Facultad de Minas. Universidad Nacional de Colombia, 2011.

[12] Saaty, T.L., *The Analytic Hierarchy Process*. McGraw Hill, Nueva York, 1980.

[13] Romero, C., *Análisis de las Decisiones Multicriterio*, Madrid, Spain: Isdefe, 1996.

ENERGY COMPARISON BETWEEN DIFFERENT PARALLEL HYBRID VEHICLES ARCHITECTURES

D. LANZAROTTO, M. PASSALACQUA & M. REPETTO
University of Genoa, Italy.

ABSTRACT

A great boom of hybrid vehicles has taken place on the automotive market in recent years, in particular, all these vehicles are now equipped with a continuously variable transmission (CVT) thanks to the use of a planetary gear train and two electric motor-generators.

The benefit provided by this system is the possibility to optimally control the engine velocity from an energy standpoint; in addition, drive comfort is increased thanks to the continuously variable transmission.

However, this is obtained at the cost of some amount of electrical losses in the components necessary to realize the above-mentioned structure.

This paper aims to evaluate the overall efficiency of this particular power train on different road missions; the same missions will be simulated at the same time for an identical hybrid vehicle equipped with a conventional transmission system.

In order to perform an energy analysis of the two architectures, one has to accurately address the main components generating energy losses: it will be thus presented the set of equations from which the mathematical stationary model of the CVT was obtained and how the different electric components and the internal combustion engine were modeled.

In addition, a brief description on the CVT optimization logic will be reported, the validity of this process will be then confirmed by comparing the ICE working points deriving from it and those declared by Toyota.

Finally, the fuel economy values coming from various road simulations will be compared in order to determine if or which hybrid architecture proves to be the most efficient one.

Keywords: Continuously variable transmission, Hybrid vehicle, Parallel architecture, Powertrain efficiency

1 INTRODUCTION

The current state of art in the automotive sector sees the presence of hybrid vehicles equipped with a continuously variable transmission (CVT) realized with a planetary gear train and two electrical machines (A. Mahmoudi, W. L. Soong, G. Pellegrino and E. Armando). This CVT is made possible by the motor-generator unit connected to the sun gear (MGU_1); this component enables the internal combustion engine (ICE), connected to the carrier gear, to rotate at the most suitable velocity.

MGU_1 can work both as a generator and as a motor depending on the transmission ratios realized, thus power is exchanged with the electrical machine connected to the ring gear (MGU_2) by a fixed-speed reduction gear.

In addition, the storage system allows regenerative deceleration and pure electric traction.

The CVT makes possible an optimal management of the ICE velocity; however, in order to work it creates electrical power circulation between the two motor generator units with its consequent losses.

It is thus of primary importance to analyze the electrical machines and their inverter efficiency because of the fundamental role of the electric components. This study was conducted in (T. A. Burress, C. L. Coomer, S. L. Campbell, L. E. Seiber, L. D. Marlino, R. H. Staunton, and J. P. Cunningham) for a 2007 Toyota Camry, in (T. A. Burress, C. L. Coomer, S. L. Campbell, A. A. Wereszczak, J. P. Cunningham, L. D. Marlino,) for a 2008 Toyota Lexus and in (T. A. Burress,

DOI: 10.2495/EQ-V2-N4-370-380

S. L. Campbell, C. L. Coomer, C. W. Ayers, A. A. Wereszczak, J. P. Cunningham, L. D. Marlino, L. E. Seiber, H. T. Lin) for a 2010 Toyota Prius.

Discontinuously variable transmission (DVT) is the alternative to the above-mentioned structure, this configuration is not able to optimally manage the ICE working conditions as well as the previous structure does; however, this transmission solution is free from any electrical losses which are only associated with regenerative deceleration and electric traction.

The aim of this paper is to evaluate the power train energy efficiency over different working conditions in order to point out the most virtuous solution from an energy efficiency standpoint.

2 HYBRID POWERTRAIN

A logical scheme of the two architectures taken into consideration is shown in Fig. 1, the two inverters are included in the 'inverter block' (only in the CVT solution) and the DC/DC converter related to the battery.

In order to briefly present the optimization process regarding the choice of the optimal transmission ratio according to road variables, the power flow scheme of CVT architecture is shown in Fig. 2 (the power flows are to be considered positive if in accordance with the arrow direction).

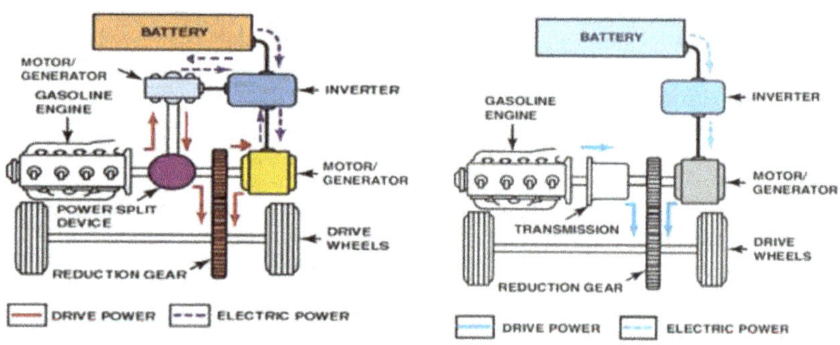

Figure 1: CVT (left) and DVT (right) chitectures.

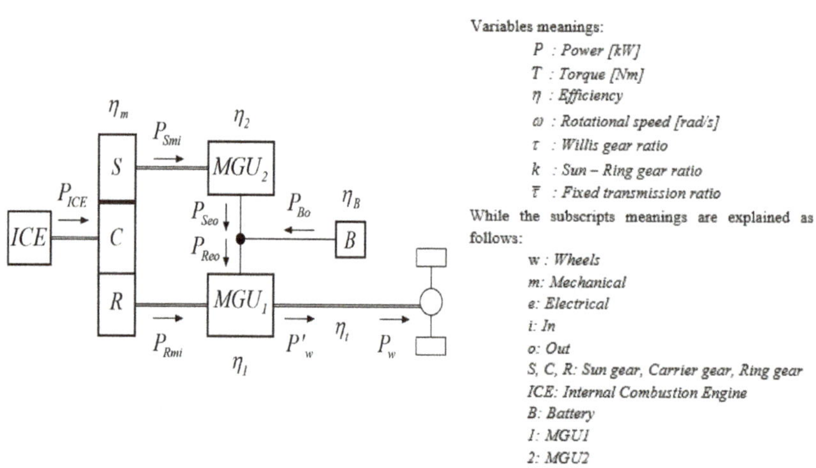

Figure 2: Power flow scheme of CVT architecture.

The analysis carried out in this study, due to the limited influence of the inertial terms of the planetary gear train components on the overall efficiency, does not consider these terms. Thus, the equations shown below are only algebraic equations and are to be considered as a stationary model.

$$\tau = \frac{\omega_R - \omega_{ICE}}{\omega_S - \omega_{ICE}}$$

where:

$$\tau = \frac{z_S}{z_R}$$

z_S and z_R represent the sun and ring teeth number, in particular z_R is conventionally a negative number because the teeth are internal.

In this study, z_S and z_R are chosen as, 30 and – 78, respectively.

Some fundamental relationships, which will be used in the model construction, are shown below:

$$\begin{cases} T_{ICE}\eta_m = T_{smi} + T_{Rmi} & \textit{torque balance at the joint} \\ P_{ICE}\eta_m = P_{Smi} + P_{Rmi} & \textit{power balance at the joint} \\ \left(1 - \frac{1}{\tau}\right)\omega_{ICE} = \omega_S - \frac{1}{\tau}\omega_R & \textit{Willis formula} \end{cases}$$

By conveniently combining the above equations, it results:

$$T_S = -\tau T_R$$

From which:

$$T_{ICE} = \frac{\left(1 - \frac{1}{\tau}\right)}{\eta_m} T_S$$

On the basis of the three following equations, once the value of k has been chosen, it is possible to obtain a relationship between the mechanical ring gear power and the mechanical sun gear power.

$$\begin{cases} T_S = -\tau T_R \\ P_{Smi} = \omega_S T_S \\ P_{Rmi} = \omega_R T_R = -\frac{\omega_R}{\tau} T_s \end{cases} \rightarrow \frac{P_{Smi}}{P_{Rmi}} = -\tau k$$

where:

$$k = \frac{\omega_S}{\omega_R}$$

Two different mathematical models of the vehicles described were implemented on MATLAB Simulink.

3 ICE AND ELECTRIC COMPONENT EFFICIENCY

It is of primary importance, as previously stated, to properly address the losses in the main components in order to effectively compare the two architectures. One referred to (G. Pellegrino, A. Vagati, P. Guglielmi and B. Boazzo) for losses modeling of the motor-generators (Fig. 3 (G. Pellegrino, A. Vagati, P. Guglielmi and B. Boazzo)).

Torque, speed and losses were normalized with the aim of referring these values to the same type of electric machines (interior permanent magnet synchronous motor) characterized by similar, but not equal, rated power as proposed by (G. Pellegrino, A. Vagati, P. Guglielmi and B. Boazzo).

Thanks to this procedure, the electrical loss contour map in Fig. 3, once normalized, can be used both for MGU_1 and MGU_2.

The efficiency map related to the inverter presented in (Hyeokjin Kim, Hua Chen, Jianglin Zhu, Dragan Maksimovic and Robert Erickson) is shown in Fig. 4. From these values, one can compute the consequent losses and their corresponding values for a 650 DC link voltage by interpolation.

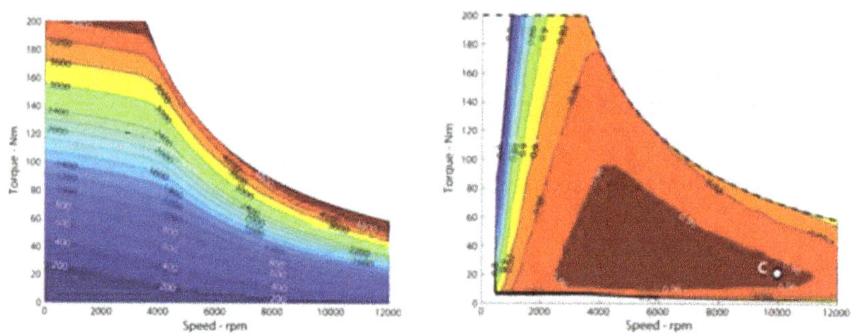

Figure 3: Efficiency and loss contour map for a 73 kW interior permanent magnet synchronous motor *(G. Pellegrino, A. Vagati, P. Guglielmi and B. Boazzo).*

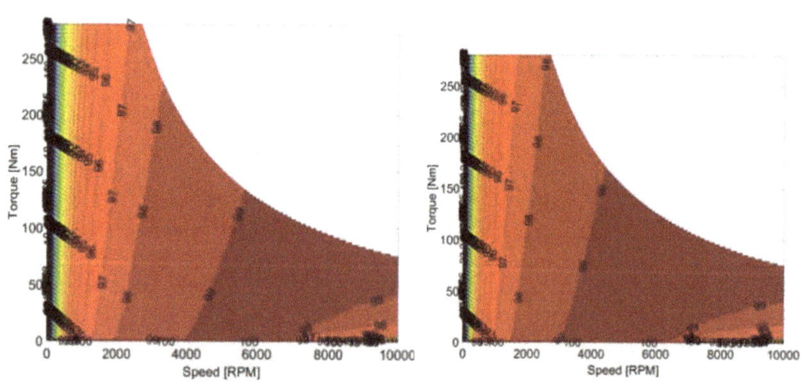

Figure 4: Sic Inverter efficiency for a 350 DC voltage and 800 DC voltage, respectively *(Hyeokjin Kim, Hua Chen, Jianglin Zhu, Dragan Maksimovic and Robert Erickson).*

Therefore, the approximating loss functions are:

$$\begin{cases} \text{Losses}_{[W]} = 2.5T_{[Nm]}^{1.18} & for \quad n_{[rpm]} < 6000 \\ \text{Losses}_{[W]} = 2.5T_{[Nm]}^{1.18} + 350\left(\dfrac{n_{[rpm]} - 6000}{4000}\right)^2 & for \quad n_{[rpm]} > 6000 \end{cases}$$

The losses in the battery were estimated from the experimental data available at the mechanical engineering department of the University of Genoa (DIME). As far as the DC/DC converter (necessary to connect the battery to the DC link) is concerned, a constant efficiency value equal to 0.97 has been considered.

Accurately evaluating the losses in this component would need an in-depth study; however, this is not relevant to the objectives of this study. As a matter of fact, the power flowing out and in the storage system is the same, and both of the structures analyzed have the same DC/DC converter, hence the same losses associated with each. In addition, given the architecture taken in consideration (parallel), large variation of these two components efficiency slightly affects the overall fuel economy.

On balance, the comparison between the two architectures would not be altered by any errors in the modelling of either the DC/DC converter or the storage system. Finally, as regards the internal combustion engine (ICE), a 72 kW petrol engine was chosen, its efficiency contour map is shown in Fig. 5 (Achim konigstein, Uwe Dieter Grebe, Ko Jen Wu, Per-Inge Larsson - Differenzierte analyse von downsizing- konzepten M.T.Z., 06/2008).

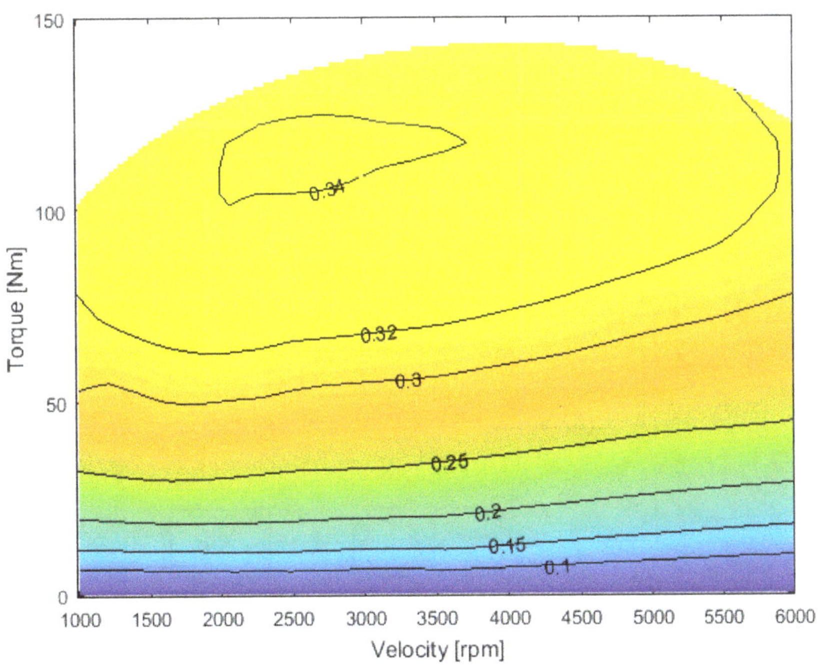

Figure 5: ICE efficiency map *(Achim konigstein, Uwe Dieter Grebe, Ko Jen Wu, Per-Inge Larsson - Differenzierte analyse von downsizing- konzepten M.T.Z., 06/2008).*

4 CONTROL LOGIC AND CVT OPTIMIZATION

The aim of the control logic implemented in the vehicle models is to instantaneously reduce fuel consumption in accordance with that presented in (Jinming Liu, Huei Peng).

This logic is simple if one looks at the DVT architecture, in that the goal is to maximize the ICE efficiency in every working condition. On the other hand, the CVT optimization process also has to take into account also the electrical losses caused by the power circulation between the two motor generator units.

A logical block scheme shows in Fig. 6 how to achieve the optimal transmission ratio k_{opt} for each couple of torque and speed (where $k_{opt} = \omega_S / \omega_R$).

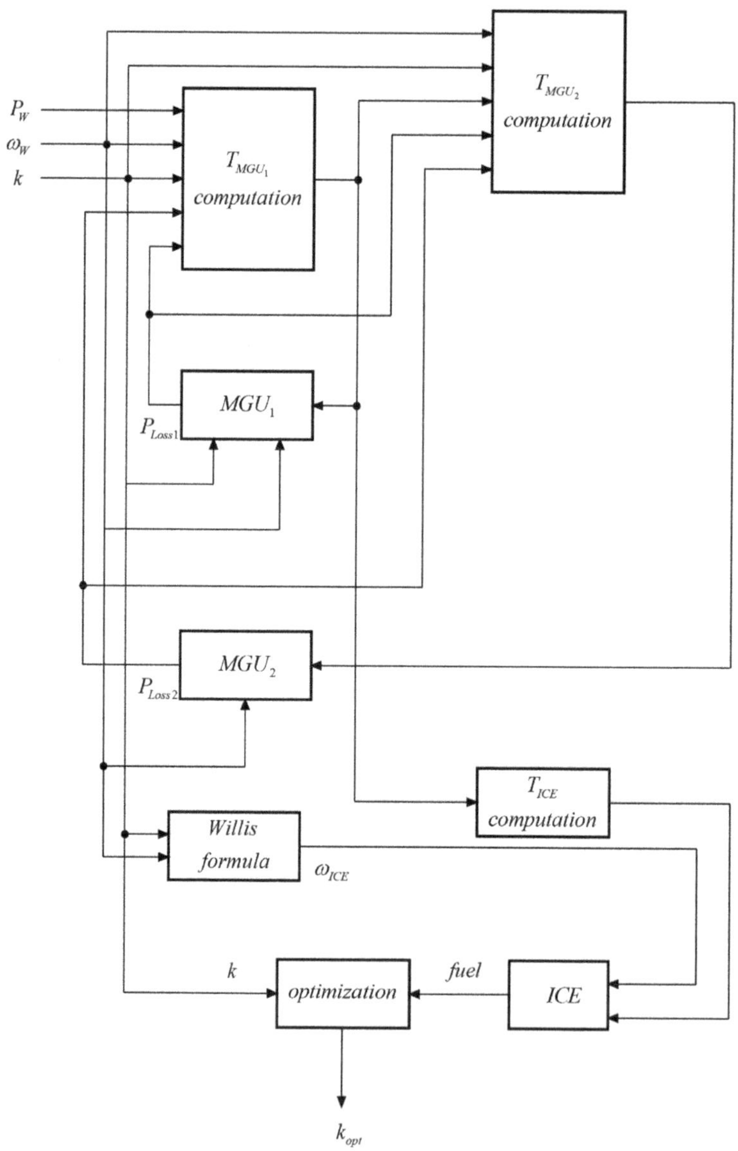

Figure 6: CVT optimization process.

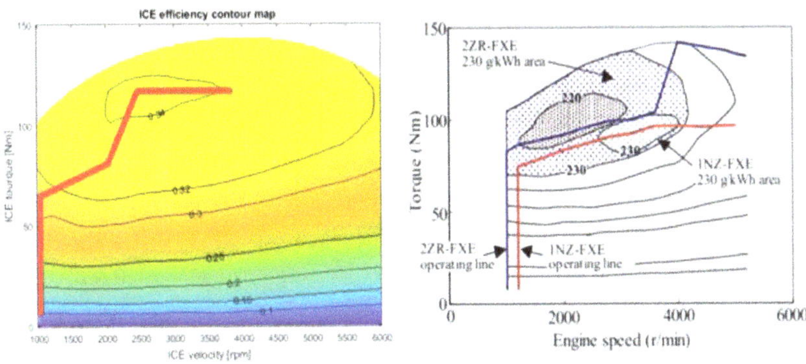

Figure 7: ICE working point deriving from CVT optimization (left), ICE working points declared by Toyota.

The effect of the optimization process regarding the choice of k_{opt} for each couple of torque and speed are visible in Fig. 7 where the ICE working points are highlighted with a red line. In particular, this result agrees with that presented by Toyota (Shouji ADACHI, Hikomasa HAGIHARA).

The transition from electric traction to ICE traction is a function of the state of charge (SOC) of the storage system; besides, it was chosen to charge the battery only through regenerative deceleration. All these logics were implemented the same way on the two vehicles in order to perform a valid comparison.

5 VEHICLE PARAMETERS AND ROAD MISSIONS

The US06, UDDS and HWFET cycles allow an immediate comparison of the results obtained in this study with those already available in the scientific literature but their drawback is given by being characterized by constant slope.

Therefore, three extra missions were simulated, whose data were experimentally acquired; two of them take place in the city of Genoa (urban mission and fast urban mission) and the third takes place around Genoa in the countryside (extraurban mission).

The main parameters related to all the missions simulated are listed in Table 1, while the main feature of the vehicles taken into consideration are reported in Table 2.

Table 1: Road mission main parameters.

	Average speed [kph]	Maximum speed [kph]	Length [km]	Change of altitude[1] [m]	Maximum road slope [%]
US06	78	130	13	-	-
UDDS	31	90	12	-	-
HWFET	78	97	16.5	-	-
Urban	24	57	11.4	20	negligible
Fast urban	27	68	22	62	6
Extra urban	45	80	36	300	9

[1]The starting altitude is the same as the final altitude, thus the positive change of altitude is intended which equals then the negative.

Table 2: Vehicle data.

Mass [kg]	1450
Rolling coefficient	0.01
Aerodynamic coefficient	0.25
Front section [m^2]	2.3
Wheel radius [m]	0.3
Final gear ratio	3.45/10.8
Final gear ratio efficiency	0.97
Planetary gear train/DVT efficiency	0.95/0.97
Air density[kg/m^3]	1.22
ICE maximum power [kW]	72
ICE maximum torque [Nm]	142
MGU_2 maximum torque [Nm]	163
MGU_2 base speed[2] [rpm]	3000
MGU_2 maximum speed [rpm]	17000
MGU_1 maximum torque [Nm]	43
MG_1 base speed [rpm]	5000
MG_1 maximum speed 10000 [rpm]	1000
DC link voltage [V]	650

6 SIMULATION RESULTS

Observing the fuel consumption results over the different missions simulated would be sufficient to identify the most efficient solution; however, in order to understand the motivation that lies beneath, one must also take into consideration the average ICE efficiency and the electrical losses.

The ICE working points are represented both for CVT and DVT architectures, in particular the points in Fig. 8 are given by urban mission and the points in Fig. 9 represent extra urban mission.

From a quick look, it is easy to notice that the DVT not being able to continuously vary the transmission ratio forces the ICE to work in a broader area even in zones distant from the optimal efficiency line. On the contrary, this line is easily followed by the CVT architecture.

Consequently, the average ICE efficiency is 3% higher in the CVT system than in the DVT structure as reported in Table 3. Despite this fact, the electrical losses due to the realization of the transmission ratio (electrical losses due to electric traction and regenerative deceleration are not included) cancel the benefit given by the thermal generation efficiency increase: the amount of these losses goes from 20% to 13% (depending on the mission type) of the energy generated by the engine.

Thus, thanks to the CVT structure, the engine is able to work at higher efficiency values, but the electrical losses deriving from the electrical power necessary to make this operation possible penalize the overall power train energy efficiency.

The fuel economy, average ICE efficiency and electrical loss values deriving from the missions simulated are reported in Table 3.

[2]Note that the DVT architecture is characterized by the presence of only one electric motor, here supposed to be identical to MGU2.

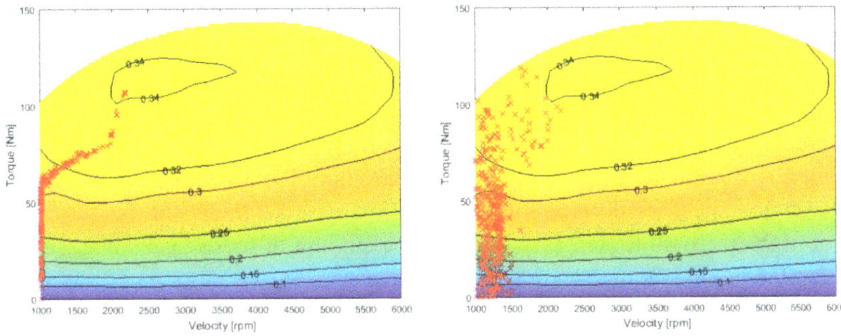

Figure 8: Urban mission ICE working points: CVT (left), DVT (right).

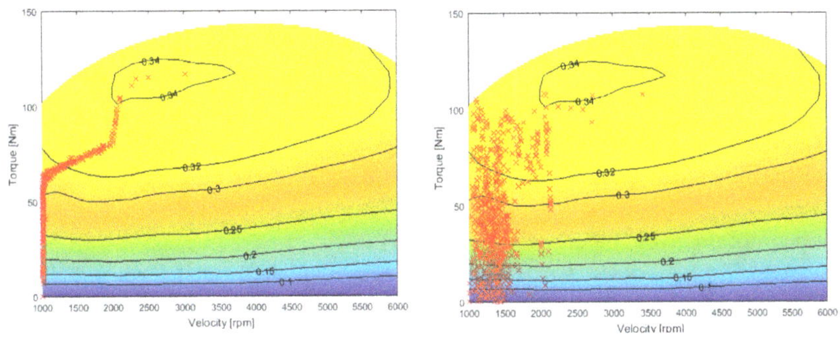

Figure 9: Extra urban mission ICE working points: CVT (left), DVT (right).

Table 3: Fuel economy, ICE efficiency and electrical losses on different missions

	Fuel economy [kpl]		Average ICE efficiency		Electrical losses[3] [%]	
	CVT	DVT	CVT	DVT	CVT	DVT
US06	15.7	17.4	0.34	0.305	17	-
UDDS	21.3	23	0.3	0.27	15	-
HWFET	21.6	23.7	0.32	0.27	21	-
Urban	24.5	25.3	0.28	0.25	13	-
Fast urban	25	26.6	0.3	0.275	13	-
Extra-urban	26.7	29.3	0.3	0.27	15	-

It is easy to notice that DVT structures allow a fuel saving from 5% to 10% compared to CVT solutions. In particular, the difference in fuel economy is higher in high-speed missions (highways), where CVT transmission in order to optimize the overall efficiency needs a superior current to flow from MGU_2 to MGU_1 and thus generating anyway higher electrical losses.

[3]The electrical losses here considered are only the ones associated with the realization of the variable transmission (they are therefore absent in the other architecture). Their amount is given in percentage with respect to the ICE generated energy.

This increase in electrical losses during high-speed situations can be found in Table 3 looking at US06 and HWFET results.

7 CONCLUSIONS

Two different hybrid parallel architectures were analyzed and their mathematical models were created and implemented on MATLAB Simulink. The various road mission simulations showed how the CVT architecture is able to increase the thermal efficiency but only at the cost of high electrical losses; in particular, the amount of these losses is not sufficiently balanced by the first advantage achieved.

It is interesting to notice that electrical losses associated with the continuously variable transmission are significantly higher in high-speed mission, and in particular, their entity is comparable to the electrical losses due to deceleration energy recovery plus those related to electric traction.

On the basis of this study, even if it provides a higher driving comfort, architectures based on the continuously variable transmission (currently the best-selling technology) do not seem to represent the optimal energy solution.

REFERENCES

[1] Mahmoudi, A., Soong, W.L., Pellegrino, G. & Armando, E., Efficiency maps of electrical machines. *2015 IEEE Energy Conversion Congress and Exposition (ECCE)*, pp. 2791–2799, 2015.
https://doi.org/10.1109/ecce.2015.7310051

[2] Burress, T.A., Coomer, C.L., Campbell, S.L., Seiber, L.E., Marlino, L.D., Staunton, R.H. & Cunningham, J.P., *Evaluation of the 2007 Toyota Camry Hybrid Synergy Drive System*, ORNL/TM- 2007/190, UT-Battelle, Oak Ridge National Laboratory, Oak Ridge, Tennessee, 2008.

[3] Burress, T.A., Coomer, C.L., Campbell, S.L., Wereszczak, A.A., Cunningham, J.P. & Marlino,L.D., *Evaluation of the 2008 Lexus LS 600h Hybrid Synergy Drive System*, ORNL/TM-2008/185, UT-Battelle, Oak Ridge National Laboratory, Oak Ridge, Tennessee, 2009.

[4] Burress, T.A., Campbell, S.L., Coomer, C.L., Ayers, C.W., Wereszczak, A.A., Cunningham, J.P., Marlino, L.D., Seiber, L.E. & Lin, H.T., *Evaluation of the 2010 Toyota Prius hybrid synergy drive system*. Oak Ridge National Laboratory (ORNL); Power Electronics and Electric Machinery Research Facility, Technical Report ORNL/TM2010/253.

[5] Pellegrino, G., Vagati, A., Guglielmi, P. & Boazzo, B., Performance comparison between surface-mounted and interior PM motor drives for electric vehicle application. *IEEE Transactions on Industrial Electronics*, **59**(2), pp. 803–811, 2012.
https://doi.org/10.1109/tie.2011.2151825

[6] Kim, H., Chen, H., Zhu, J., Maksimovic, D. & Erickson, R., Impact of 1.2kV SiC-MOSFET EV traction inverter on urban driving. *2016 IEEE 4th Workshop on Wide Bandgap Power Devices and Applications (WiPDA), Fayetteville, AR*, pp. 78–83, 2016.

[7] konigstein, A., Grebe, U.D., Wu, K.J. & Larsson, P.-I., Differenzierte analyse von downsizing- konzepten. MTZ - Motortechnische Zeitschrift, **69**(6), pp. 468–476, 2008.
https://doi.org/10.1007/bf03227458

[8] Liu, J. & Peng, H., *Control Optimization for a Power-Split Hybrid Vehicle*. Proceedings of the 2006 American Control Conference, Minneapolis, Minnesota, USA, 2006.

[9] Adachi, S. & Hagihara, H., TOYOTA MOTOR CORPORATION, AICHI JAPAN The renewed 4-Cylinder Engine Series for Toyota Hybrid System.

[10] Liu, J., Peng, H., & Filipi, Z, *Modeling and Control Analysis of Toyota Hybrid System*. Proceedings, 2005 IEEE/ASME International Conference on Advanced Intelligent Mechatronics, pp. 134–139, 2005.

OPTIMISING CAR LIFE FOR MINIMUM CO_2 EMISSION

HARRY C. WATSON
School of Engineering, University of Melbourne, Australia.

ABSTRACT

In this paper, the historical trends and future projections of whole of life CO_2 emissions is followed and includes the changing effects on embedded production energy as vehicles have been made lighter. Even so, the rapid reduction in fuel consumption of conventional vehicles leads to the ratio of embedded to in-use CO_2-e to have doubled in the last 30 years. This embedded energy sourced CO_2 recurs each time a new car is made, so the front end energy has to be amortised over the life of the vehicle. It is shown that the ratio is several times higher for battery electric vehicles, while hybrids fall between electric and conventional. The importance of vehicle useful life is emphasized. In the past, the optimum life to amortise the embedded energy was about 17 years but this depends on the prevailing rate of improvement in in-use energy of the marketed fleet. The paper concludes on the basis of the evidence presented that the optimum life for present conventional vehicles is between 10 and 12 years and for battery electric vehicles approaching 20 years with hybrids falling between. As the rate of annual fuel consumption improvement reduces from the present level of 5%/y, the desirable life-times of vehicles will increase. It is recommended that some form of government policy be implemented to achieve the changes in optimum vehicle life-time, over the next few decades, through support for 'Cash for clunkers' or equivalent mechanisms. This will enable the most rapid achievement of greenhouse gas emissions reduction. Incentives or other mechanisms need to be found to encourage hybrids rather than all electric vehicles to achieve best possible vehicle fleet CO_2 reduction.

Keywords: Lifetime CO_2, embedded energy, in-use energy, conventional engines, hybrid, all electric, market trends, policy outcomes

1 INTRODUCTION

This paper is founded on research done by the author for SAE-A's submission to a government committee looking to set Australia's long-term fuel consumption standards. The author was, in 2004, responsible for advising the government on the 2010 CO_2 standards for all light duty vehicles [1, 2]. In that work, the lowest attainable CO_2 emission at reasonable cost was determined. However, no consideration was given to any impacts of the non-fuel CO2 contributions to whole of life emissions. The non-fuel-related CO_2 emissions have grown from 16% in 1980 [3], but despite initiating policies to reduce manufacturing energy consumption [4], it had grown to 20% by1997 [5] largely as the result of reduced vehicle fuel consumption in meeting government targets caused by perceived energy shortage and the 'energy crisis' initiated by the oil shocks of 1973 and 1979. Although attempts have been made to include manufacturing energy as a vehicle design objective, only small reductions in embedded energy appear to be possible[6].

Thus, today it now represents about 30% of total CO_2 assignable to light duty vehicles and therefore cannot be ignored [7]. This proportion will increase as the fuel efficiency of vehicles continues to improve more rapidly than the remaining energy inputs (sometimes with an increase in manufacturing energy as with hybrids).

2 OBJECT

The object of this paper is to compare the fixed and variable energy content in passenger car life cycles and to determine what is the optimum service life to minimise greenhouse gas emissions. The problem is treated for the EU market, although the method will be applicable to other markets.

DOI: 10.2495/EQ-V2-N4-381-388

3 METHOD

For this investigation, it is assumed that the only CO_2-producing components are the manufacturing energy and embedded materials energy offset by recycled materials and parts, together with the in-service fuel and consumables sourced CO2-e. It is worth noting that the in-service consumables other than fuel are of second-order importance and their variation in consumption over the vehicle life or vehicle technology is assumed to be constant irrespective of age [5]. The assumption is made that the fuel consumption (use per distance travelled) does not vary as the vehicle ages [7].

Thus, the problem reduces to finding

$$CO2_{\min} = \min[\sum_{y=1}^{n} CO2_{fuel}(y_1).y + CO2_{embeded}\ ]_{c=1}^{a}. \tag{1}$$

where $CO2_{fuel}(y)$ is part of a series in CO_2 or fuel consumption that for new cars progressively that reduces (e.g. 3%/y) as manufacturers meet the requirements of reduced fuel consumption for compliance with EU regulations.

n is the number of years of service life before it is replaced by a new vehicle.
and a is the number of life cycles c studied

Accounted for in modelling is that in the early years of use vehicles typically travel more than in later life as found in the ABS SMVU [10] This variation shown in Fig. 1 is quite consistent with the mean travel and age of the EU population [8, 9]. Subjectively, this can be explained as many new cars are used for business purposes early in their lives, and as second or third hand cars used less for only local trips late in their lives when reliability and other defects make them less attractive for long journeys. Moreover, these data represent an average car, and thus include scrapping (retiring) cars from the fleet through crashes and then later owner decisions to send the car to recycling, which significantly increases after the median age.

The relative importance of the 'front end' manufacturing and materials energy or embedded energy (and corresponding CO_2-e emissions) can be expressed as a ratio to the average annual energy/emissions from fuel use. This variation is given in Fig. 2. It can be seen that between 1978 and 2012 embedded CO_2-e rose from 15% to 31% in the whole of life analysis. During this time, the overall life cycle CO_2-e fell by just over 40% for the class-E-sized

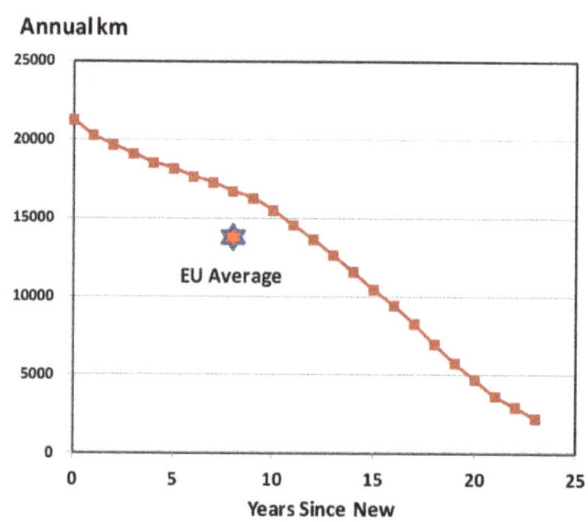

Figure 1: Annual km of travel in Australia, [7] compared with European data [8, 9].

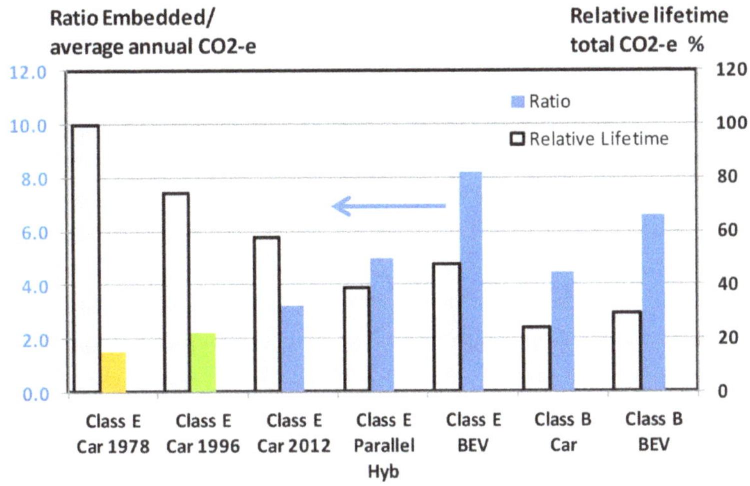

Figure 2: Embedded CO_2-e compared with annual use and relative CO_2-e for life cycle with a base of 1978 [3] orange and 1996 [6] green. Blue values 2013 [11].

car. For the Battery Electric Vehicle (BEV), there is a further 10% reduction, but the hybrid of the day had another 20% reduction or at 40% of the baseline 1978 car. Results for 2012 class B size conventional and BEV cars are given. Note worthy are the relatively high ratios of embedded versus annual use sourced CO_2-e for the BEV cars because of the significant emissions from battery production.

3.1 Example small improvement

A graphical example of the application of equation 1 is given in Fig. 3. To illustrate the concepts in the example, it is assumed that the initial CO_2 is four times the first year in-service

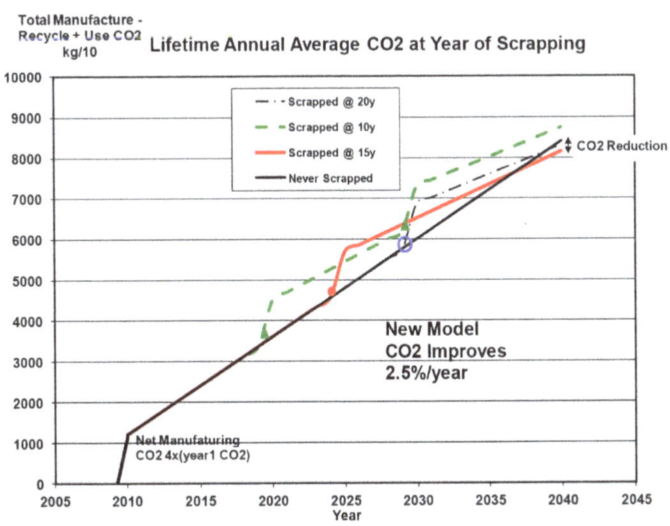

Figure 3: Example of three replacement lifetimes in a market with 2.5%/y reduction in fuel consumption and embedded energy/CO_2-e of four times first year fuel use.

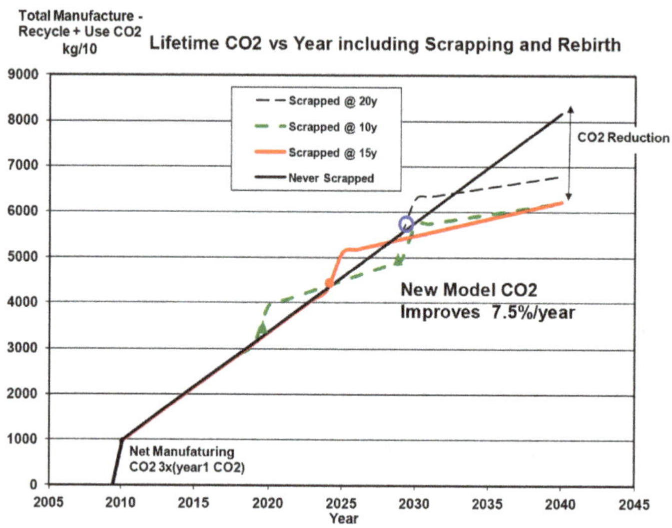

Figure 4: Example of three replacement lifetimes in a market with 7.5%/y reduction in fuel consumption and embedded energy/CO2-e of 3 times first year fuel use.

CO_2 (or equal to 4 years of fuel use). A constant average travel per year is selected (e.g. assumed to be 15,000 km/y). The vehicle chosen emits 200 g/km CO_2 at the beginning and throughout its life. This vehicle is sold into a market where replacement vehicles have a 2.5%/y improvement in fuel consumption year by year, so any replacement vehicle will have lower fuel consumption; its reduction depends on its time of delivery into the market. From the three lifetimes illustrated, the best CO_2-e reduction for somewhere between 15 and 20 y lifetimes. The CO_2 mitigated is only a few per cent.

3.2 Example of large improvement

This vehicle is now sold into a market where replacement vehicles have a 7.5%/y improvement in fuel consumption and embedded energy of three times in the first year use. Best CO_2-e emission reduction is between from a 10 and 15 y life to scrap with around 25% reduction in CO_2-e from continuous use of the original vehicle, after 30 y as seen in Fig. 4.

4 MARKET TRENDS AND NEEDS

Figure 5 shows the EU trend in fleet average fuel consumption [12] and the projection to meet the 2020 target. It is apparent that the existing trend, changing at about 4.5% reduction in fuel consumption and therefore CO_2-e emissions per year, if maintained will meet the EU target of 95 g/km. Likely, industry will achieve lower because manufacturers have to individually meet the target and will need to do so with a safety margin. This is illustrated by the fleet average of about 120 versus the 130 g/km CO_2 required for 2015. Therefore, it is highly likely that an improvement rate close to 5% will continue.

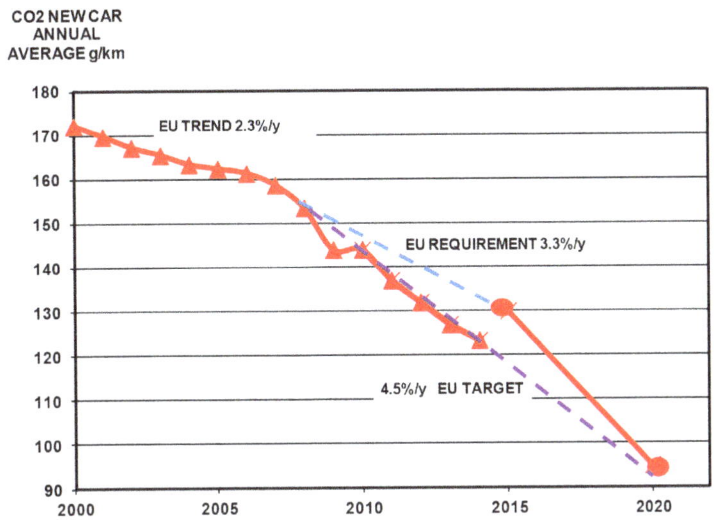

Figure 5: European new car annual average CO_2 emissions and targets.

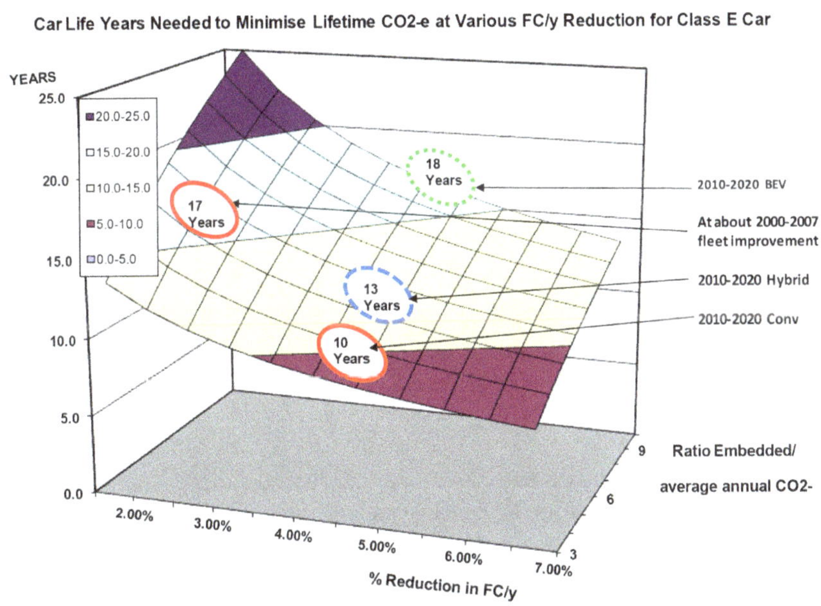

Figure 6: Optimum vehicle life for class E cars.

5 RESULTS

5.1 Results for class E cars

Modelling of a class E car has been done using eqn 1 with the variable km of travel per annum given if Fig. 1 and the ratio of embedded to year 1 emission of CO_2-e for the range seen in Fig. 2. The results, in Fig. 6, show the surface for minimum CO_2-e as it varies with this ratio and the annual rate of improvement in fuel consumption (and CO_2-e) to give the

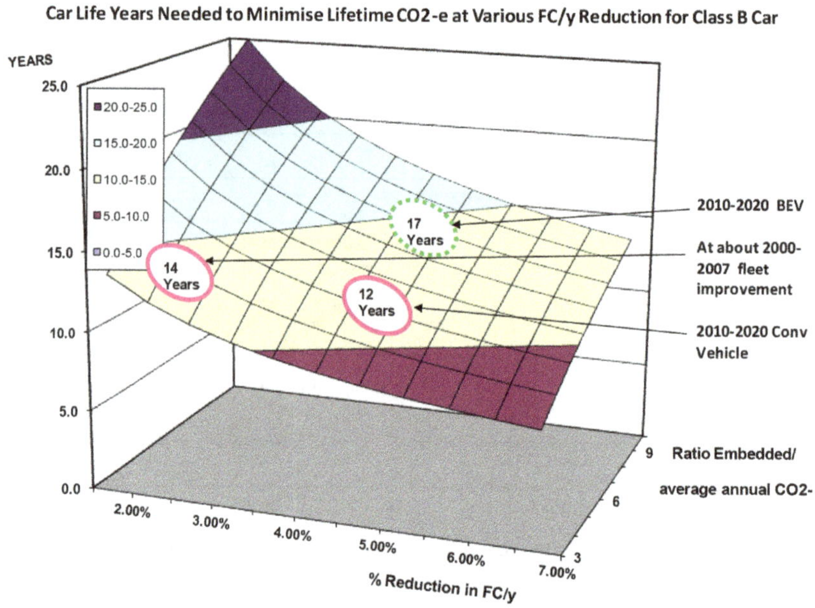

Figure 7: Optimum vehicle life for class B cars.

corresponding optimum vehicle service life. It can be seen that for the low rate of fuel consumption reduction in the period prior to 2007 the optimum vehicle life was around 17 years, whereas for current cars with a nearly 5% improvement in fuel consumption per year the optimum life reduces to 10 years.

Based on the assumption that hybrids and battery electric vehicles (BEV) improve at the same rate as the market a whole, their optimum life is longer 12 and 18 years, respectively, because of higher embedded energy (and CO_2-e) lower energy consumption in service. If their improvement rate is different, the surface allows alternative conclusions to be drawn.

5.2 Results for class B cars

The results for the smaller class B cars show a similar shaped surface with optimum lives of 14 years for older vehicles and for 12 and 17 years as the optimum for current and BEVs, respectively.

6 DISCUSSION

Inclusion of the embedded materials and production energy, with allowance for recycling, significantly delays the time of benefit from introducing a new fuel efficient car fleet. Unsurprisingly, the lower the in-use energy consumption (lowest for BEVs), the longer the cars need to operate in service to amortise their high embedded energy through their service life, which is a consequence of the low energy density of the battery and their relatively higher mass, even though more weight saving technologies are found in BEVs. The longer optimum life suggested for BEVs and hybrids leads to the question of whether these batteries have the durability to last perhaps twice the normal warranted battery life of ten years. Also for BEVs, the well-proven assumption, that in-service energy consumption does not vary with age, may not be valid as battery efficiency deteriorates with age as does battery available kWh. This is in contrast to the parallel hybrid, which in Fig. 2, already has the lowest lifetime energy consumption and is optimum when retained for a service life of 13 years as seen in Fig. 6 for class E cars.

It can be hypothesised that as the law of diminishing returns takes effect, that the per cent rate of improvement in fuel consumption will reduce, and that the optimum life will move back to higher years of in-service. In the market place, the resale price of the Toyota Prius has held up at least as well as the conventional C class cars, indicating that relatively less demanding use compared with a BEV may enable battery replacement free lives of the high teen years. This along with the prospects of significantly increased improvement possible in the internal combustion engine as found in Formula 1 racing cars for example make a strong case for the encouragement of mild hybrid uptake in the market.

7 CONCLUSIONS

It has been the province of automobile clubs to give advice to members on how long cars and vans might be used for life-time minimum per km travel cost; often slated as about 17 years. A similar approach has been applied to determine the optimum life of vehicles to minimise their life-time CO_2-e when their manufacturing and materials energy (including recycling) is added to their life-time fuel use emissions. If the life-time is short, there will be a need to enact legislation and incentives to terminate the life of a vehicle. 'Cash for clunkers' was a marketing ploy in the past. It is concluded:

1. The results here suggest for the European market, with the needed rapid improvement in vehicle fuel consumption to meet the EU 2020 target CO_2 emission regulation, that normal E and B class vehicles have an optimum life now of about 10 and 12 years, respectively. Mechanically many of these vehicles could continue to operate through to 15 to 20 years. However, as the rate of fuel consumption reduction reduces in the future, as it must, the optimum life will increase perhaps to 17 to 20 years.
2. It is recommended that some form of government policy be implemented to enable this process, of changing vehicle life-time, in the decades ahead in support of the most rapid achievement of greenhouse gas emissions reduction.
3. If battery electric vehicles (BEV) follow similar usage patterns to regular cars, their optimum life needs to be longer at the present time, at around 18 years, in a market with 5% improvement energy consumption per year. If their annual travel is less than that of the fleet average, as it is likely to be, the time in service should be even longer. This seems to be in conflict with early adopters and fleet buyers who are major purchasers of BEV cars, since the whole of car BEV image is important and the support for interior/infotainment systems etc in the very long term is unlikely. Therefore, it may be anticipated that owners are likely to be early rather than late scrapping their cars.
4. On the other hand, hybrids, even the simpler parallel hybrid used as the example here, looks to be the outstanding technology with an optimum life-time in the present market of 13 years and likely increasing to high teen years as technological improvements diminish.
5. These findings indicate that governments should subsidise hybrids, instead of the present support for all electrics i.e. BEVs.
6. Finally, it is obvious that these findings depend on the energy mixes in vehicle production and materials, and the proportion of CO_2-free electricity produced. Here, these values are based on the best available information. If vehicle production involved no CO_2-e

emission, these findings would be overturned. However, it is very unlikely that major components of the car such as steel, tyres etc will be made without carbon emissions.

ACKNOWLEDGEMENT

The author wishes to acknowledge support from the Centre for Automotive Research and Testing (ACART) and the co-operation of colleagues Profs Michael Brear and Chris Manzie.

REFERENCES

[1] ACIL Tasman and Watson H.C., *Setting the passenger car fuel consumption target for 2010,* Report prepared for FCAI and Australian Greenhouse Office, 1999. .

[2] ACIL Tasman and Watson H.C., *National average carbon dioxide emissions by vehicles, possible target for new light duty vehicles in 2010,* Report prepared for FCAI and Australian Greenhouse Office, 2004. .

[3] Watson, H.C., *Passenger car life cycle energy consumption.* S.A.E.-A. Paper No. 79313, pp. 1–10, 1979.

[4] Watson, H.C., *Summary of S.A.E.-Australia's Energy Policy.* S.A.E.-A. Paper No. 79310, pp. 1–10, 1979.

[5] Parikh Y. & Watson, H.C., *Life cycle emissions manufacturing/ use/infrastructure.* In Transport Engine Emissions, Advanced Engineering Centre for Manufacturing, University of Melbourne, Victoria, 1997.

[6] Watson, H.C., Charters, W.W.S., Brey, S., Parikh, Y., Lamb, D.G. & Fewchuk, D., *Concept Car - Life Cycle Energy Analysis.* SAE paper 981154, 1998.

[7] Beer, T., Grant, T., Watson, H.C. & Olaru, D., *Life-Cycle Emissions Analysis of Fuels for Light Vehicles* Report (HA93A-C837/1/F5.2E) to the Australian Greenhouse Office , 2004.

[8] Schipper, L. *Automobile Fuel Economy and CO2 Emissions in Industrialized Countries: Troubling Trends through 2005/6* . EMBARQ, the World Resources Institute Center for Sustainable Transport, Washington, 2008.

[9] Available at: https://setis.ec.europa.eu/sites/default/files/reports/Driving_and_parking_patterns_of_European_car_drivers-a_mobility_survey.pdf

[10] ABS SMVU Australian Bureau of Statistics, Triennial Surveys of Motor Vehicle Usage , 1976.

[11] Sharma, R., Manzie, C., Breeside, M., Brear, M.J. & Crawford, R.H., Conventional, hybrid and electric vehicles in Australian driving conditions - Part 2 Life cycle CO2-e emissions. *Transportation Research Part C*, **28**, pp. 63–67, 2013.

[12] Available at: http://www.theicct.org/sites/default/files/publications/ICCTbriefing_EUCO2_201507.pdf. (accessed April 2017).

EXPLORING THE INTEGRATION OF THE POWER TO GAS TECHNOLOGIES AND THE SUSTAINABLE TRANSPORT

E. LLERA[1], L. M. ROMEO[1], M. BAILERA[1] & J. L. OSORIO[2]
[1] Escuela de Ingeniería y Arquitectura, Departamento de Ingeniería Mecánica, Universidad de Zaragoza, Campus Río Ebro, María de Luna, Zaragoza, Spain.
[2] CIRCE Institute -Universidad de Zaragoza, CIRCE Building – Campus Río Ebro, Mariano Esquillor Gómez, Zaragoza, Spain.

ABSTRACT

The de-carbonization of the transport sector is a particularly complex challenge as greenhouse gases are delocalized and diffused. Therefore, the problem has to be tackled from the source of the emissions, and efforts in the scientific and technological field must seek out new energy vectors of high density, neutral in CO_2 and based on renewable energy that meet the sector demands and requisites. This could be the case of the synthetic natural gas which can be produced through the Power to Gas process (PtG). This process, originally developed by the German institutes ZSW and IWES, converts electricity into synthetic natural gas (SNG) via the methanation of CO_2 together with H_2 from water electrolysis. The energy content of the produced methane comes from the primary source for power generation (optimally renewable electricity) and it is possible to produce a CO_2 neutral fuel by capturing the carbon emissions from an existing source. In addition, the PtG process can be seen as a new concept of renewable energy and CO_2 hybrid storage. This paper identifies the possibilities that the Power to Gas technology offers for the production of sustainable methane and the existing potential for the symbiosis of industrial sectors through optimization of their waste streams of matter and energy. In particular power and transport sectors are considered and the outline of a small facility for the generation of synthetic natural gas from renewable electricity and its consumption in the vehicles of a road freight company is presented as a case study. Not only the technical feasibility but the economic viability of the process and the environmental improvements resulting from the use of a renewable fuel free of CO_2 emissions in terms of carbon footprint are evaluated.

Keywords: alternative fuels, carbon capture, energy storage, Power-to-gas, synthetic natural gas.

1 INTRODUCTION

In order to improve the security of the European energy supply and reduce the emissions of greenhouse gases (GHG), the European Union (EU) set in 2001 the objective of replacing 20% of conventional fuels by 2020 through the introduction of liquid biofuels in the short term, natural gas (NG) in the medium and long term and hydrogen in the distant long term [1]. In addition, it was expected that biofuels shared at least 6% of the fuels used in road transport in 2010. However, biofuels only shared 4.4% by 2010 [2], which was mainly the result of socioeconomic problems generated in Europe and in developing countries by the production of feedstocks, such as an increase in food prices and land use competition [3]. For that reason, the European Commission (2013) recommended different alternative fuels based on the maturity of the technologies for each application, such as electricity, compressed natural gas (CNG) and hydrogen, for urban use vehicles and liquefied natural gas (LNG) for long-haul transport.

It is broadly agreed that electric vehicles powered by renewable energy vectors (battery electric vehicles (BEVs) for renewable electricity or fuel cell electric vehicles (FCEVs) for renewable hydrogen) can help EU towards a more sustainable transport system and meeting its goal to reduce greenhouse gas emissions by 80–95% by 2050 [4]. However, larger numbers of electric vehicles will not be enough for the shift to a low-carbon economy and in the short and medium term, it is clear that Europe will still have to rely on conventional internal combustion engine vehicles (ICEVs), while new and cleaner technologies develop.

DOI: 10.2495/EQ-V3-N1-1-9

In the meanwhile, the natural gas can play a bridging role as it can be used both in spark-ignition and in compression-ignition engines [5]. NG has been widely and profitably applied in compressed form in many countries mainly in urban vehicles. CNG has only been popular in light-duty vehicles due to the low energy density of gaseous NG, which gives low autonomy, whereas by storing NG in liquid form vehicles can increase their autonomy by up to 1100 km [6]. Recent studies suggested that LNG use in heavy duty vehicles (HDV) has the potential to reduce environmental impacts and noise in cities, in addition to the maturity of the technology, energy resource availability and clear interest of the EU in supporting the LNG adoption [7].

The use of NG in vehicles is encouraged by the benefits of reducing local air pollution, resources availability, the existence of distribution infrastructure and relatively lower prices than petroleum fuels [8].

Regarding the GHG emissions, the use of CNG in light vehicles reaches up to the 25% reduction in CO_2 equivalent emissions in its lifecycle comparing to those fed by gasoline. Fed with LNG trucks can achieve low emissions without excessive and expensive emission control equipment as required by diesel engines. The process of production of LNG requires natural gas first dried and then submitted to the removal of hydrocarbons, sulphur and CO_2, what gets a high-purity methane (98%); therefore LNG vehicles produce even lower quantities of polluting emissions than CNG vehicles [9]. This decrease is even greater when compared with diesel vehicles: 80% less carbon monoxide, 70% less NOx and 45% less unburned hydrocarbons (HC). With respect to diesel, total of CO_2 equivalent GHG emissions reduction is a 15–18%, while solid particles (compared to diesel fuel with low sulphur content), it is more than 97% [10].

Finally, the use of NG in road transport presents as an added bonus considerable economic savings, because of the lower price of natural gas when in general compared to diesel. For instance, in Spain, savings between 21% and 34% per travelled kilometre have been estimated for intercity road freight transport meaning that the extra investment for the natural gas technology investments can be amortized over the course of one year [11].

Despite the technical feasibility of the NG as transport fuel has been widely demonstrated, up to date there has been a weak growth of the use of natural gas for road transport. Three barriers have been identified: the high cost of vehicles, the low level of consumer acceptance and the scarcity of refueling stations. The clean fuel strategy launched by the European Commission [2] and the 2014/94/EU Directive [12] are aimed to encourage the use of alternative fuels by means of the installation of a network of refuelling station through Europe.

But even though the full integration of the natural gas in road transport would lead to a reduction of the environmental impacts of the sector, it would not completely solve its share of responsibility to global warming and the depletion of resources. In the medium to-long term, natural gas of fossil origin must make way to sustainable fuels.

The generation of synthetic natural gas based on renewable energy and neutral in CO_2 could be the solution.

In the last years, Power to Gas has been proposed as a very promising and versatile technology that converts electricity into synthetic natural gas via the methanation of CO_2 together with H_2 from water electrolysis [13,14]. The energy content of the produced methane comes from the primary source for power generation (renewable energy in a sustainable scenario) and it is possible to produce a fuel which does not increase net GHG emissions into the atmosphere if the required CO_2 is captured from an existing source [15, 16].

The process was originally developed by the German institutes ZSW (Zentrum für Sonnenenergie- und Wasserstoff-Forschung Baden-Württemberg), and IWES (Fraunhofer

Institute for Wind Energy and Energy System Technology) and there are currently 30 demonstrating plants in Europe [17].

In practice, both existing and projected PtG facilities integrate electrolyser with a wind or photovoltaic installation, but PtG may be used to store surplus or curtailed electricity whilst converting to a gas vector [18]. In any case, this technology would allow the connection between electric and gas network thus increasing the flexibility of the energy supply system [19].

Although this would be also true as for the hydrogen production [20], distribution and use of methane is far more readily available than hydrogen based on the current infrastructure.

Economic related barriers such as the large required investments could be overcome if the PtG process is seeing as a new concept of renewable energy and CO_2 hybrid storage. Symbiosis of industrial sectors through optimization of their waste streams of matter and energy could increase the global efficiency, improve the resource use, and reduce the investment on new equipment. Some of the synergies of these hybrid systems are already being tested under real conditions (e.g. PtG-Amine scrubbing, PtG-wastewater treatment) while others have been studied through numerical simulations for instance PtG-oxyfuel combustion or PtG-Electrochemical hybridization [21].

According the German Federal Ministry of Transport and Digital Infrastructure [22] the development of business models as a step of preparing the markets and synergy with other energy sectors is one of the priority fields of action for the deployment of the PtG in the transport sector.

This paper identifies the possibilities that PtG technology offers for the production of sustainable methane and the existing potential for the symbiosis of the power and the transport sector.

A small facility for the in-situ generation of synthetic natural gas from renewable electricity and CO_2 from transport emissions then consumed in the vehicles of a road freight company is used as base case to assess the economic viability of the gas from different electricity origins and at different production scales. In addition, the GHG reduction that could be achieved by the use of synthetic methane as a substitute for diesel transport fuel is assessed.

2 CASE STUDY

The PtG process is conceptually explained as follows: renewable electricity is converted to fuel gas by means of electrolysis, storing electrical energy in form of hydrogen. Oxygen is also obtained as by-product from the electrolyser. Then, the generated hydrogen is combined with carbon dioxide to produce methane through the Sabatier reaction:

$$CO_2 + 4H_2 \leftrightarrow CH_4 + 2H_2O \qquad \Delta H_{298K} = -165\,kJ/mol \tag{1}$$

Many studies consider the use of surplus or curtailed renewable power [14] from existing installations and the total H_2 produced is modelled based on the obtainable electricity [23].

The availability of a suitable source of CO_2 can be considered as the main limiting factor for promoting PtG, since it reduces the possible geographic sites for a wide deployment of this technology [24]. Biogas plants, waste managers, energy intensive industries and power plants are the largest CO_2 sources and the most interesting partners for integration with PtG [25]. In the transport sector, carbon capture and storage (CCS) into light vehicles has been also focused. Post-combustion-capture technologies has been identified as the most easily adapted for use with the existing fleet of internal combustion engines [26].

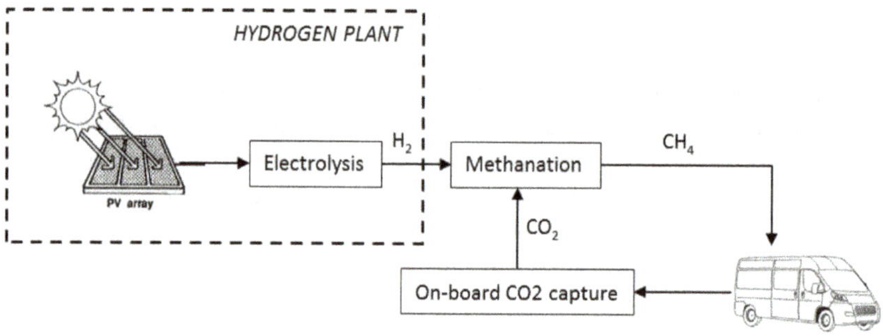

Figure 1: Scheme of the installation for SNG production.

The case under study is based on an isolated installation for the production of renewable and neutral synthetic natural gas in the headquarters of a road freight transport company in Spain.

As shown in Fig. 1, the installation consists of an electrolytic hydrogen production plant and a TREMP methanation plant where hydrogen and CO_2 react to produce synthetic methane. In the present study, an isolated PV installation is included in order to assure the renewable origin of the electricity that is consumed in the electrolyser.

Even though CO_2 could be provided by other sources, it would be more difficult to demonstrate the neutrality in terms of GHG which is one of the goals for this study. Then it is considered that CO_2 derives from the exhausted gases of the vehicles properly separated.

Finally, the produced gas is supposed to be consumed by the company's own vehicles that operate currently with fossil natural gas.

Considering Sabatier reaction and combustion reaction for the methane:

$$CH_4 + 2O_2 \rightarrow CO_2 + 2H_2O \tag{1}$$

the volume of CO_2 emitted is equal to the volume previously fixed during the methanation process.

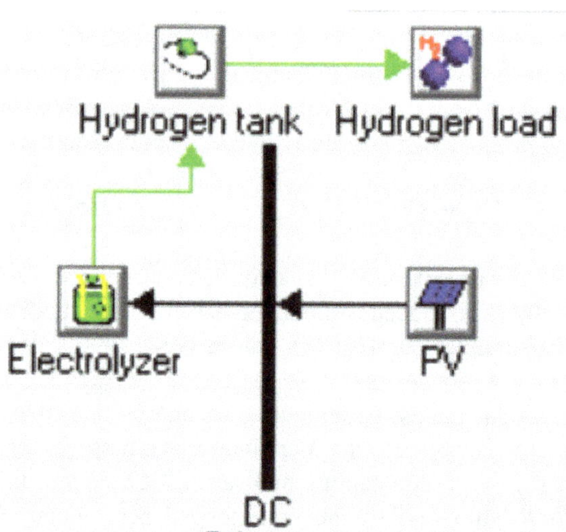

Figure 2: Schematic of the hydrogen plant.

3 METHODOLOGY AND RESULTS

The first step is analysing the activity and the fleet owned by the company in order to estimate the amount of natural gas needed to a reliable and stable delivery over time. The natural gas consumption determines the hydrogen and CO_2 demand as according the Sabatier reaction, four times of H_2 and one times of CO_2 in volume are required for the production of one volume of CH_4.

The annual consumption of SNG is estimated considering the technical specifications of the natural gas fuelled vehicles that are owned by the company. One of these vehicles consumes 4.3 kg of CNG per 100 km. Considering a 1.3 correction factor (for efficiency losses due to the CCS system), 200 km distance travelled daily and 260 labour days, the vehicle would consume 2900 kg of synthetic natural gas per year or 11.18 kg per day. It is assumed that natural gas is fuelled to the vehicle from 6 to 8 AM.

The second step is studying the viability of the hydrogen plant in the considered location which requires optimizing the size and number of the system components. As shown in Fig. 2, the size of the PV system, electrolyser and hydrogen storage tank must be determined in this case to ensure that the plant is able to meet the required hydrogen load at the lowest possible cost. The optimization software used for this work is the Hybrid Optimization Model for Electric Renewables (HOMER) from the National Renewable Energy Laboratory (NREL), which was freely available at the moment of the estimations. HOMER is a tool that implements comparative economic analyses on distributed generation power systems. Inputs to HOMER include load data, renewable resource data, system component specifications and costs, and various optimization parameters. HOMER performs an hourly simulation of every possible combination of components entered and ranks the systems according to user-specified criteria, such as cost of energy or capital costs.

To obtain the input data for HOMER, information of the hydrogen plant components was collected from research literature and manufacturers to obtain an estimation of hydrogen system costs and efficiencies. Also collected are the load and resource time series data and information necessary to model the systems.

The operating principle is to use a PV system to directly supply the electrolyser that meet the hydrogen load. When excess PV energy is available, the generated hydrogen is stored in a tank.

Hydrogen load data were estimated considering the SNG consumption and the conversion rates according the Sabatier reaction. A production of around 1454 kg of hydrogen per year or 5.59 kg per day is needed. It is assumed that hydrogen plant is operated from 7 AM to 3 PM.

Solar radiation and clearness data for the installation site were for GMT Time Zone. The annual average global radiation is 5.507 kWh/m^2/day and the annual average clearness index is 0.573.

This scenario was modelled with PV panels fixed at a slope of 30° and lifetimes were assumed to be 20 years.

In HOMER, a solution is a particular hybrid system configuration. As HOMER searches for the optimal hybrid system design, it typically evaluates a high number of solutions. HOMER performs an hourly time series simulation of every solution and eliminates solutions that cannot meet the required load. For the case under study, HOMER examined the 1080 solutions possible by combining components.

The least-cost technically viable solution consists in a PV installation of 100 kW, one electrolyser of 50 kW and compressed gas storage with 100 kg of capacity. Next Table 1 summarized the main costs of the hydrogen installation.

Table 1: Cost for the hydrogen production plant.

Initial capital	Operating cost ($/yr)	Total NPC
$ 544 206	45 982	$ 1 132 015

With a total electrolyser production of 1434 kg of hydrogen per year, the levelised cost of hydrogen is 60.9 $/kg.

Regarding the availability of CO_2 it is considered the use of those emitted by the vehicle that is around 8000 kg per year.

Onshore capture methods with amines have been tested only for their use in ships, including compression process for convenient and efficient transportation and storage. However, if compression methods are applied on engines mechanically, then serious problems arise, such as power penalties, space requirements for storage, and safety issues.

Several author have set that the energy consumption required to attain concentrated CO_2 streams leads to efficiency penalties that range between 9 and 12 efficiency points, and the associated costs may vary from 50 to 90 $/t$CO_2$, depending on CO_2 concentration in flue gas and capture technology [27]. The higher limit has been considered for economic considerations in this paper.

Methanation costs of 10 $/kg and 1.01 $/t for gas compression are considered [28].

The cost of producing 1 kg of SNG in the proposed installation is calculated as 40.4 $ that is twice as high as the market price for fossil natural gas.

Regarding CO_2 emissions, the EU Renewable Energy Directive of 2009 states that biofuel emissions are calculated as zero due to balancing the amount of carbon released with an equivalent amount sequestered, therefore such emissions were not considered. Then, the environmental benefits considered in this paper are the GHG emissions (in carbon dioxide equivalent, CO_2eq) saved when the total CH_4 produced is used to replace diesel fuel.

Lifecycle assessment results from literature were collected in order to determine GHG emissions in replacing diesel with renewable gas. The "Well to Wheel" life cycle assessment includes emissions associated with fuel production, processing, transportation, distribution and consumption.

The total CO_2eq saved when the total CH_4 produced is used to replace diesel fuel is calculated as substituted fuel in MJ multiplied by the emission factor in kgCO_2eq/MJ. Considering an emission factor of 2.61 kgCO_2eq/MJ, each vehicle saves around 10 tons of CO_2 per year.

Although this paper proposes the integration of renewable energy systems for both transport and decentralized production of fuels as a way to achieve security of supply, fuel costs are quite high but within the limits if R&D is still continuing.

4 CONCLUSIONS

This paper has approached the combination of renewable electricity and CO_2 sourced from transport emissions in order to provide a source of renewable transport fuel.

The existing potential for the integration of sectors such as energy and transport through optimization of their flows of matter and energy (industrial symbiosis) has been demonstrated.

The obtained results may open new prospects for the development of infrastructure of supply and storage and production distributed synthetic natural gas as alternative fuel, enhancing its availability and promoting their full integration in the transport sector.

In order to reduce the current costs, the search for innovative solutions that allow adapting the transport infrastructures to the efficient use of alternative energy in order to reduce exploitation costs must be pursued.

ABBREVIATIONS

BEVs: battery electric vehicles, CCS: carbon capture and storage, CNG: compressed natural gas, CO2eq: carbon dioxide equivalent, EU: European Union, FCEVs: fuel cell electric vehicles, GHG: greenhouse gases, HC: hydrocarbons, HDV: heavy duty vehicles, IWES: Fraunhofer Institute for Wind Energy and Energy System Technology, LNG: liquefied natural gas, NG: natural gas, PtG: Power to Gas, SNG: synthetic natural gas, ZSW: Zentrum für Sonnenenergie- und Wasserstoff-Forschung Baden-Württemberg

REFERENCES

[1] ETP, "White paper European transport policy for 2010: time to decide. Commission of the European communities," p. 2010, 2001.

[2] European Commission, "Alternative fuels for sustainable mobility in Europe - European Commission," 2013 [Online], available at: https://ec.europa.eu/transport/themes/urban/cpt_en (accessed 17 July 2017).

[3] EEGFTF, "Future Transport Fuels, Report of the European Expert Group on Future Transport Fuels," Düsseldorf, 2011.

[4] European Environment Agency, "Electric vehicles and the energy sector - impacts on Europe's future emissions — European Environment Agency," 2016 [Online], available at: https://www.eea.europa.eu/themes/transport/electric-vehicles/electric-vehicles-and-energy (accessed 17 July 2017).

[5] Hekkert, M.P., Hendriks, F.H.J.F., Faaij, A.P.C. & Neelis, M.L., Natural gas as an alternative to crude oil in automotive fuel chains well-to-wheel analysis and transition strategy development. *Energy Policy*, **33**(5), pp. 579–594, 2005. https://doi.org/10.1016/j.enpol.2003.08.018.

[6] DENA, "LNG in Germany: Liquefied Natural Gas and Renewable Methane in Heavy-Duty Road Transport," 2014 [Online], available at: http://www.dena.de/en/topics/energy-efficient-transport-systems.html (accessed 17 July 2017).

[7] Osorio-Tejada, J.L., Llera-Sastresa, E. & Scarpellini, S., Liquefied natural gas: could it be a reliable option for road freight transport in the EU? *Renewable and Sustainable Energy Reviews*, **71**, pp. 785–795, 2017. https://doi.org/10.1016/j.rser.2016.12.104.

[8] Yeh, S., An empirical analysis on the adoption of alternative fuel vehicles: the case of natural gas vehicles. *Energy Policy*, **35**(11), pp. 5865–5875, 2007. https://doi.org/10.1016/j.enpol.2007.06.012.

[9] Chamberlain, G.A., MANAGEMENT OF LARGE LNG HAZARDS, 2006 [Online], available at: http://citeseerx.ist.psu.edu/viewdoc/download?doi=10.1.1.567.7712&rep=rep1&type=pdf (accessed 17 July 2017).

[10] Fedorova, E.B., Fedorov, V.V. & Shakhov, A.D., Promising technology for recovery and use of liquefied natural gas cryogenic engineering, production and use of industrial gases, refrigeration engineering. *Chemical and Petroleum Engineering*, **45**(3), pp. 3–4, 2009.

[11] GASNAM, Estrategia de Impulso del vehículo con energías alternativas (VEA) en España (2014–2020), 2015.

[12] European Parliament and the Council of the European Union, "DIRECTIVE 2014/94/EU OF THE EUROPEAN PARLIAMENT AND OF THE COUNCIL - of 22 October 2014 - on the deployment of alternative fuels infrastructure -," 2014.

[13] Götz, M., Lefebvre, J., Mors, F., Koch, A.M., Graf, F., Bajohr, S., Reimert, R. & Kolb, T., Renewable Power-to-Gas: a technological and economic review. *Renewable Energy*, **85**, pp. 1371–1390, 2016. https://doi.org/10.1016/j.renene.2015.07.066

[14] Rönsch, S., Schneider, J., Matthischke, S., Schlüter, M., Götz, M., Lefebvre, J., Prabhakaran, P. & Bajohr, S., Review on methanation – From fundamentals to current projects. *Fuel*, **166**, pp. 276–296, 2016. https://doi.org/10.1016/j.fuel.2015.10.111

[15] Streibel, M., Nakaten, N., Kempka, T. & Kühn, M., Analysis of an integrated carbon cycle for storage of renewables. *Energy Procedia*, **40**, pp. 202–211, 2013. https://doi.org/10.1016/j.egypro.2013.08.024

[16] Kühn, M., Nakaten, N., Streibel, M. & Kempka, T., CO_2 geological storage and utilization for a carbon neutral "power-to-gas-to-power" cycle to even out fluctuations of renewable energy provision. *Energy Procedia*, **63**, pp. 8044–8049, 2014. https://doi.org/10.1016/j.egypro.2014.11.841

[17] Bailera, M., Lisbona, P., Romeo, L.M. & Espatolero, S., Power to gas projects review: lab, pilot and demo plants for storing renewable energy and CO_2. *Renewable and Sustainable Energy Reviews*, **69**, pp. 292–312, 2017. https://doi.org/10.1016/j.rser.2016.11.130

[18] Estermann, T., Newborough, M. & Sterner, M., Power-to-gas systems for absorbing excess solar power in electricity distribution networks. *International Journal of Hydrogen Energy*, **41**(32), pp. 13950–13959, 2016.

[19] Qadrdan, M., Abeysekera, M., Chaudry, M., Wu, J. & Jenkins, N., Role of power-to-gas in an integrated gas and electricity system in Great Britain. *International Journal of Hydrogen Energy*, **40**(17), pp. 5763–5775, 2015. https://doi.org/10.1016/j.ijhydene.2015.03.004

[20] Emonts, B., Schiebahn, S., Görner, K., Lindenberger, D., Markewitz, P., Merten, F. & Stolten, D., Re-energizing energy supply: electrolytically-produced hydrogen as a flexible energy storage medium and fuel for road transport. *Journal of Power Sources*, **342**, pp. 320–326, 2017. https://doi.org/10.1016/j.jpowsour.2016.12.073

[21] Bailera, M., Espatolero, S., Lisbona, P. & Romeo, L.M., Power to gas-electrochemical industry hybrid systems: a case study. *Applied Energy*, **202**, pp. 435–446, 2017. https://doi.org/10.1016/j.apenergy.2017.05.177

[22] Bünger, U., Landinger, H., Pschorr-Schoberer, E., Schmidt, P. & Weindorf, W., Power-to-Gas (PtG) in transport Status quo and perspectives for development. *Naumann (dbfz); A. Lischke (DLR) Munich*, 2014. [Online]. Available at: http://www.lbst.de/ressources/docs2014/mks-studie-ptg-transport-status-quo-and-perspectives-for-development.pdf. (accessed 17 July 2017).

[23] Vo, T.T.Q., Xia, A., Wall, D.M. & Murphy, J.D., Use of surplus wind electricity in Ireland to produce compressed renewable gaseous transport fuel through biological power to gas systems. *Renewable Energy*, **105**, pp. 495–504, 2017. https://doi.org/10.1016/j.renene.2016.12.084

[24] Schneider, L. & Kö, E., The geographic potential of Power-to-Gas in a German model region - Trier-Amprion 5. *Journal of Energy Storage*, **1**, pp. 1–6, 2015. https://doi.org/10.1016/j.est.2015.03.001

[25] Sterner, M., Institut für Solare Energieversorgungstechnik., and Institut für Elektrische Energietechnik Rationelle Energiewandlung., *Bioenergy and renewable power methane in integrated 100% renewable energy systems: limiting global warming by transforming energy systems*. Kassel University Press, 2009.

[26] Sullivan, J.M. & Sivak, M., *Carbon capture in vehicles: a review of general support, available mechanisms, and consumer acceptance issues*. University of Michigan Transportation Research Institute, Ann Arbor, 2012.

[27] IPCC, *IPCC Special Report on Carbon Dioxide Capture and Storage. Prepared by Working Group III of the Intergovernmental Panel on Climate Change* [Metz, B., O. Davidson, H. C. de Coninck, M. Loos, and L. A. Meyer (eds.)]. Cambridge University Press, Cambridge, 2005.

[28] E&E Consultant Environement Energie, "Etude portant sur l'hydrogène et la méthanation comme procédé de valorisation de l'électricité excédentaire," 2014. [Online]. Available at: http://www.grtgaz.com/fileadmin/engagements/documents/fr/Power-to-Gas-etude-ADEME-GRTgaz-GrDF-complete.pdf (accessed 17 July 2017).

3D SHAPE OPTIMISATION OF A LOW-PRESSURE TURBINE STAGE

L. WITANOWSKI, P. KLONOWICZ & P. LAMPART
The Szewalski Institute of Fluid-Flow Machinery, Polish Academy of Sciences, Poland.

ABSTRACT

The possibility of reducing the flow losses in low-pressure turbine stage has been investigated in an iterative process using a novel hybrid optimisation algorithm. Values of the maximised objective function that is isentropic efficiency are found from 3D RANS computation of the flowpath geometry, which was being changed during the optimisation process. To secure the global flow conditions, the constraints have been imposed on the mass flow rate and reaction. Among the optimised parameters are stator and rotor twist angles, stator sweep and lean, both straight and compound. Blade profiles remained unchanged during the optimisation. A new hybrid stochastic-deterministic algorithm was used for the optimisation of the flowpath. In the proposed algorithm, the bat algorithm was combined with the direct search method of Nelder-Mead in order to refine the best obtained solution from the standard bat algorithm. The method was tested on a wide variety of well-known test functions. Also, the results of the optimisation of the other stochastic and deterministic methods were compared and discussed. The optimisation gives new 3D-stage designs with increased efficiency comparing to the original design.

Keywords: CFD, efficiency, hybrid method, low pressure steam turbine, optimisation, swarm intelligence.

1 INTRODUCTION

The performance and profitability of energetic systems depend fundamentally on turbine characteristics – the lower the turbine efficiency, the lower the system efficiency e.g. steam power plant efficiency [1, 2], ORC efficiency [3], energy storage efficiency [4]. Consequently, system efficiency basically depends on the expander design. The computational fluid dynamics (CFD) tools are extremely useful in the design of high-efficiency turbines because of inaccuracy of empirical loss correlation models [5]. The most important component of the total loss is usually connected with the losses within the blade passages – mainly profile loss and secondary (endwall) loss. Despite the significant amount of models, none of them by far has proven to be really general [6]. This leads to the use of computational fluid dynamics in an intensive way for turbomachinery blade design, but this higher performance must be achieved with fast prototyping and lower cost. The largest computational power cost during efficiency optimisation of turbine is connected with the calculation of values of the objective function, which is obtained from a RANS solver.

Novel methods must be used in order to obtain the global minimum with lower number of calls of the objective function. Among these methods are neural network metamodels [7, 8], Kriging and co-Kriging methods [9] and hybrid stochastic-deterministic algorithms [2, 10–14].

In this work, the new hybrid stochastic-deterministic optimisation method was used. The method is a combination of Bat algorithm [15] and Nelder-Mead method [16]. Also, a parametric stator and rotor geometry model has been developed. As shown in this paper, these methods allow one for efficient optimisation of turbine blading system.

DOI: 10.2495/EQ-V3-N1-10-21

2 TURBINE MODEL

The five-stage LP part of a 360 MW steam turbine was presented in Fig. 1. The LP part processes pressure drop from 514.76 kPa to 8.855 kPa, inlet temperature is equal to 537.72 K, rotational speed is 3000 rpm, steam mass flow rate is 103 kg/s. The optimisation process was focused on the first stage of the turbine (Fig. 2). The first stage consists of 202 stator and 227 rotor blades. Both have cylindrical geometry.

3 OPTIMISATION ALGORITHM

3.1 Bat algorithm

Bat algorithm is an optimisation algorithm inspired by echolocation phenomenon of bats and was proposed by Yang in 2010 [15]. Bats use a type of sonar, the echolocation tool, to recognize the prey, avoid obstacles and locate their roosting cervices in the dark [15, 17, 18]. They

Figure 1: LP part of the steam turbine.

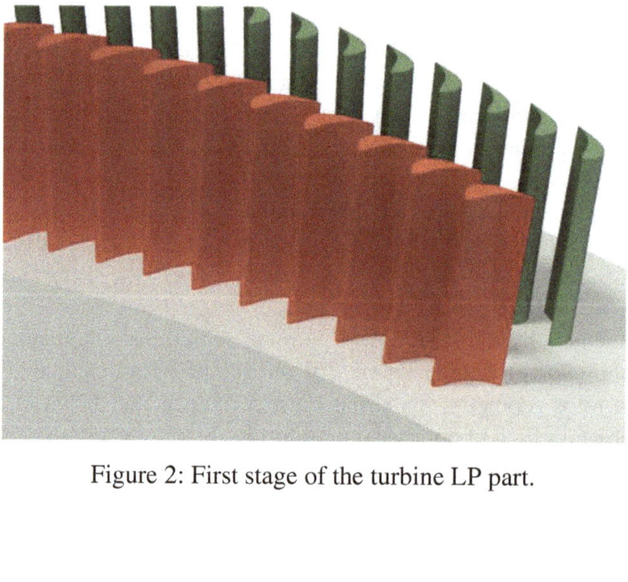

Figure 2: First stage of the turbine LP part.

emit a very loud sound pulse and listen for the echo that bounces back from the surrounding objects. Their pulse varies in properties and can be correlated with their hunting strategies, depending on the species. Their signal bandwidth varies depending on the species and often increases by using more harmonics [19].

The echolocation characteristics are idealized within the framework of the following rules by benefiting such features of bats [17]:

- All bats use echolocation to sense distance, and they also know the difference between food/prey and background barriers in a seemingly 'magical' way.
- Bats fly randomly with velocity v_i at position x_i with a fixed frequency f_{min}, varying wavelength λ and loudness A_0 to search for a prey. They can automatically adjust the wavelength (or frequency) of their emitted pulses and adjust the rate of pulse emission r in the range of [0, 1], depending on the proximity of their target.
- Although the loudness can vary in many ways, we assume that the loudness varies from a large (positive) A_0 to a minimum constant value A_{min}.

The original bat algorithm is illustrated in Fig. 3.

3.2 Nelder-Mead method

The local direct search method called the Nelder-Mead method [16] is one of the most popular nonlinear optimisation methods. This method was based on the earlier work of Spendley, Hext, and Himsworth [20]. In Nelder-Mead method for a function of n parameters, the algorithm maintains a set of n+1 function values and was evaluated at n+1 points in parameter space. This set of points defines a simplex in n dimensions. In two dimensions, the simplex would be a triangle. The worst vertex, where f(x,y) is the largest, is rejected and replaced with a new vertex. A new triangle is formed and the search is continued. The process generates a sequence of triangles (which might have different shapes), for which the function values at the vertices get smaller and smaller. The size of the triangles is reduced and the coordinates of the minimum point are found. To define a complete Nelder-Mead method, four parameters must be specified: reflection, expansion, contraction and shrinkage.

```
Initialize the bat population x_i (I = 7,2, ...,n) and velocity v_i
Define pulse frequency f_i at x_i
Initialize pulse emission rate r_i and loudness A_i
for t=1:Max number of iterations
Generate new solutions by adjusting frequency and updating velocities and location:
if rand > r_i
Select a solution among the best solutions
Generate new local solution around the selected best solution
end
Generate new solutions by flying randomly
if rand < A_i & f(x_i) < f(x_best)
Accept the new solutions
Increase r_i and reduce A_i
end
Rank the bats and find the current best x_best
end
Postprocess results and visualization
```

Figure 3: Pseudo-code of bat algorithm.

3.3 A novel hybrid algorithm

The hybridization is a connection of two or more algorithms that run together and complement each other to produce a profitable synergy from their integration [11, 21]. Hybridization aims to combine the advantages of each algorithm, whilst reducing the disadvantages. In hybrids, one algorithm may be incorporated as a sub-algorithm to locate the optimal parameters for another program or different components of algorithm such as mutation crossover that are used to improve another algorithm in the hybrid structure. The hybrid algorithms can be divided into two groups – collaborative hybrids and integrative hybrids. Depending on the combination of the running algorithm, weight of participating of the collaborative hybrids can be grouped as follows: multi-stage, sequential, and parallel.

The proposed Multi-stage Hybrid Bat Algorithm consists of Bat Algorithm and simplex Nelder-Mead method. In standard version of the bat algorithm, the local search of the algorithm is performed only by the solution to satisfy the condition written in Fig. 3 in line 6. Thus, the local search ability of the algorithm depends on a random number compared with the emission rate. The value of the pulse emission rate increases as an iteration proceeds. In order to improve local search, the Nelder-Mead method was used. The steps of new algorithm are started after line 15. Then the simplex method is run if standard bat algorithm did not improve the best solution. In this modification, all solutions are collected and are ranked before the start of the Nelder-Mead algorithm.

4 METHODOLOGY

4.1 Parametrization

The parametrization of 3D axial turbine stage was based on several changing parameters: stator twist angle, rotor twist angle, stator simple circumferential lean, rotor simple circumferential lean, stator compound circumferential lean, rotor compound circumferential lean, stator simple axial lean, rotor simple axial lean, stator compound axial lean, rotor compound axial lean. The total number of optimisation parameters was 12: 6 for stator and 6 for rotor. Each parameter was allowed to vary in a prescribed range (bounded optimisation).

4.1.1 Blade twist angle

One of the basic method of changing 3D geometry of the blade is twisting of the stator/rotor profiles. An example of this kind of optimisation can be found in literature [22]. This type of solution is often applied for the rotor blades of the MP and LP turbine parts as the flow parameters vary significantly across the blade height. With this approach, the channel throats can be decreased near the endwalls, while they can be increased in the middle section of the blade. The rotor profiles are being rotated with respect to their mass center. In case of the stator profiles, the rotation is performed about the trailing edges. Figure 4 presents the optimized rotor blade along with its initial twisting.

4.1.2 Simple and compound circumferential lean

The circumferential lean is one of the solutions reducing the variation of the flow parameters and reaction along the channel height of the LP part. The reduction of the reaction variation, increasing it near the hub and decreasing it near the shroud, leads to reduction of the supersonic flow loss in the boundary layer near the trailing edges. As a consequence, the blade loading near the hub is increased and the separation is reduced. Additionally, the blade tip

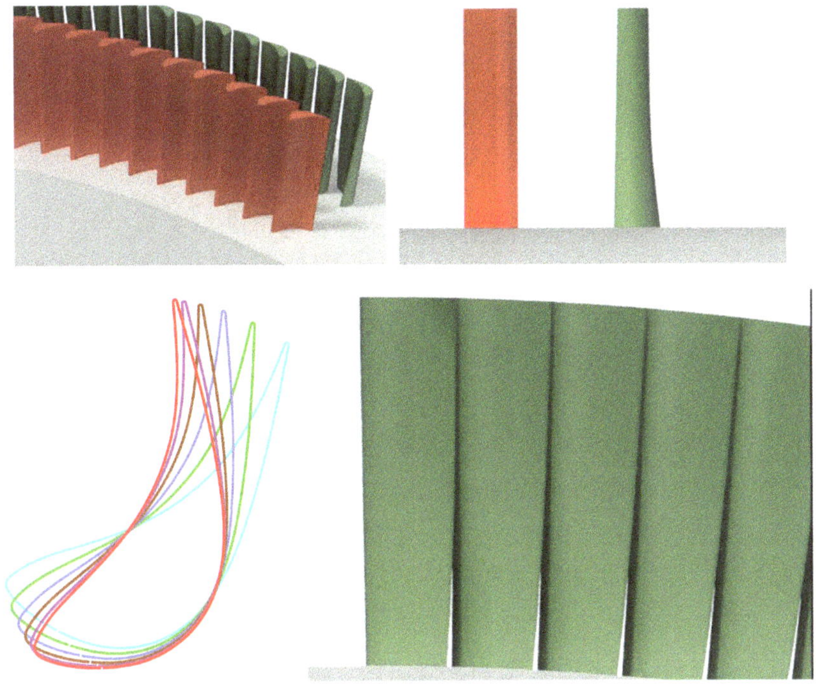

Figure 4: Rotor blade and its twisting.

leakage is reduced. The possibilities of increasing the efficiency depend on the actual loading which in case of the last stages can change in relatively wide range. The best results are obtained in the range of low loadings. Also, the variation of the stator outflow angle is reduced along the blade height which is beneficial from the point of view of the rotor blade for which the twisting level can be decreased [23].

The comparison of the original stage geometry, stage with rotor simple lean and rotor compound lean has been presented in Figs. 5–7.

Figure 5: The front view of the original rotor.

Figure 6: The front view of the rotor with simple circumferential lean.

Figure 7: The front view of the rotor with compound circumferential lean.

4.1.3 Simple and compound axial lean

The idea of the axial lean is similar to the circumferential lean; however, different effects can be obtained. For the compound lean stator blade, the midspan section move into a region of low pressure. As a consequence, the loading of blade increase [24].

The comparison of the original stage geometry, stage with rotor simple lean and rotor compound lean has been presented in Figs. 8–10.

4.2 Objective function and constrains

The isentropic total-to-total efficiency was selected as the objective function for the shape optimisation of the stator and rotor blades. With this definition, the exit kinetic energy losses are not included in general losses of the stage. In order to secure global flow conditions, constraint was imposed on the reaction degree, mass flow rate.

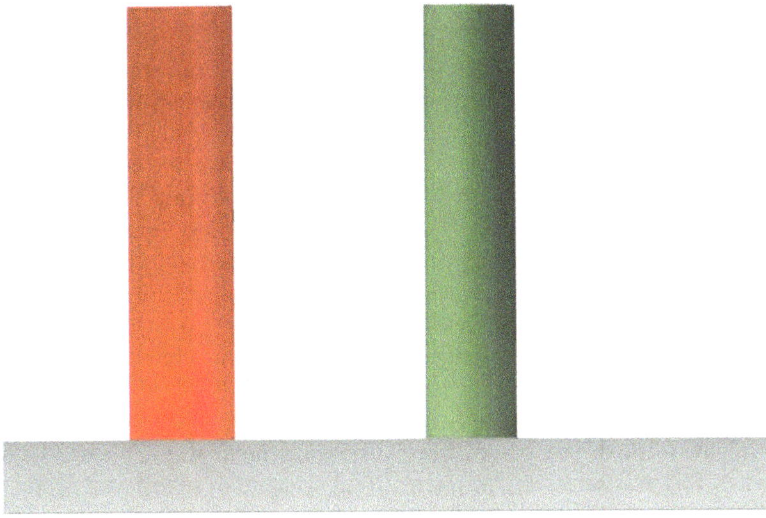

Figure 8: The circumferential view of the original stage.

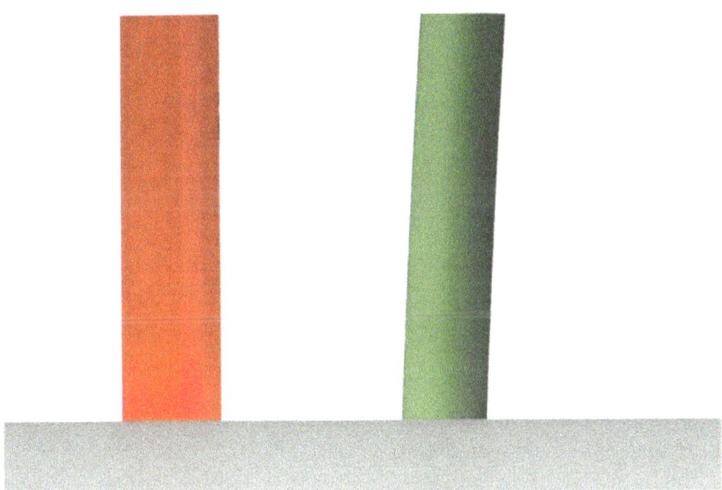

Figure 9: The circumferential view of the stage with simple axial lean of the rotor.

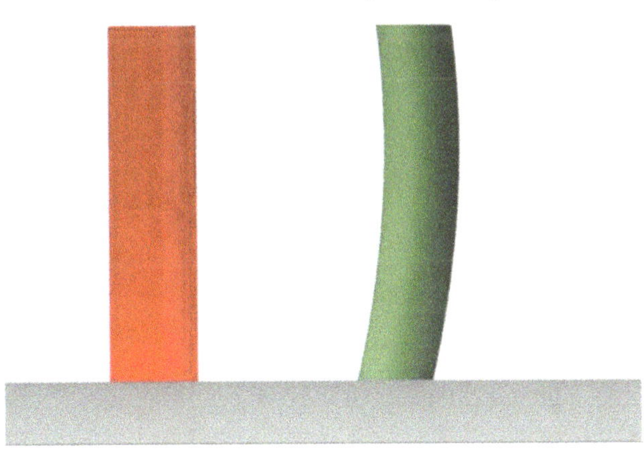

Figure 10: The circumferential view of the stage with compound axial lean of the rotor.

4.3 Procedure

The overall optimisation process consisted of two phases – iterative phase and verification. All stages of the iteration loop were managed by the script written in Matlab application. The geometry of the optimized stage was being created within the Matlab script. In order to perform the CFD analysis, the next step was the generation of suitable computational grid. An automatic grid was created by means of Ansys Turbogrid software. The hexahedral meshes prepared for the numerical simulations consisted of about 200 thousand nodes for every domain (single stator channel and single rotor channel with periodic condition). Special care was given to the edge length ratio, face angle, element volume ratio. During the optimisation, the quality of the meshes was monitored.

The RANS simulations were performed by means of a commercial code Ansys CFX. The numerical model used in the computations was similar to that used by other authors who simulated turbine stages [25, 26]. The equations describing the flow were solved by means of the finite volume method. Second-order discretization scheme was applied. As a turbulence model the k-w SST was selected. The international standard of state equations for water and steam IAPWS-IF98 was used (version for industrial calculations). The imposed set of boundary conditions consisted of the total pressure and total temperature at the inlet, average static pressure at the outlet and the rotational speed of the rotor domain.

In the verification phase, the optimal geometry was tested on a fine grid (2 mln nodes for stage). The comparative results for the original and optimized geometries presented in the paper are those for the refined grids. Three algorithms were used in order to study the optimisation process: Nelder-Mead method [27], Implicit Filtering [28], Hybrid Bat Algorithm. The optimisation scheme was presented in Fig. 12.

5 RESULTS

The optimizer chose the following changes of the original geometry of the considered LP stage:

- Stator twist angle increased 0.1875°.
- Stator simple axial lean angle increased 4°.
- Stator simple circumferential lean angle increased 2°.
- Stator compound axial lean increased 0.875.

Figure 11: Fine mesh for verification process.

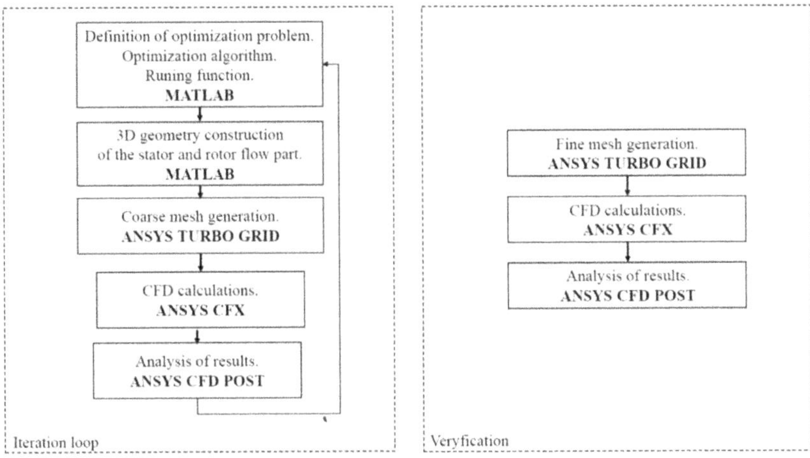

Figure 12: Optimisation strategy description.

- Stator compound circumferential lean increased 1.75.
- Rotor twist angle decreased 0.1875° and 0.125°.
- Rotor simple axial lean angle increased 4°.
- Rotor simple circumferential lean angle decreased 2°.
- Rotor compound axial lean increased 0.8125.
- Rotor compound circumferential lean decreased 2.

As a result of the optimisation process new geometry has been created. The efficiency of optimised flow path increased by about 0.4% from 96.5% to 96.9% (Fig. 13). The best value of objective function was obtained from proposed hybrid method but also other methods resulted in loss reduction. The final geometry was presented in Fig. 14.

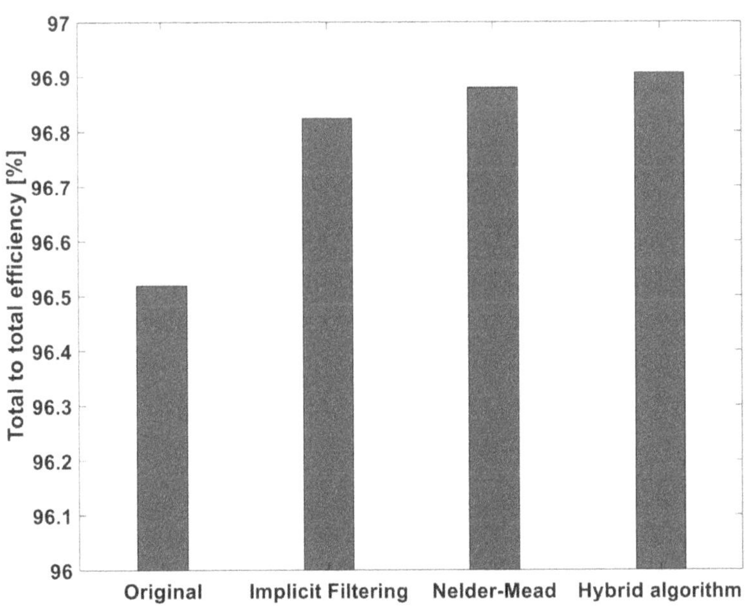

Figure 13: Comparison of total to total efficiency of original and optimized geometries.

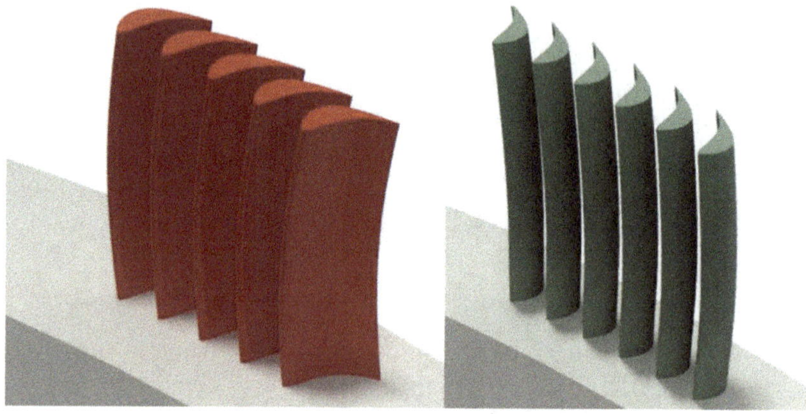

Figure 14: Optimized geometry of LP turbine.

6 CONCLUSION

The axial stage was optimized by means of the hybrid optimisation method using the combination of Bat algorithm with Nelder-Mead algorithm. The objective function – the total-to-total isentropic stage efficiency – was obtained from 3D RANS simulations. As the decision variables, the geometrical blade parameters for stator and rotor were selected. They influenced the twisting angle distribution as well as simple and compound circumferential and axial lean.

The optimization of the geometry of the low-pressure axial turbine stage resulted in noticeable efficiency increase equal to 0.4 percentage point. Similar procedure applied for the whole LP turbine part could presumably increase the whole turbine efficiency.

The further work will include new optimisation algorithms. Also, the other optimization objective function will be tested and validated.

ACKNOWLEDGEMENTS

This work was supported by The National Science Centre, Grant No. 2015/17/N/ST8/01782.

REFERENCES

[1] Lampart, P. & Yershov, S., 3D Shape optimiastion of turbomachinery blading. *TASK*, **6**(1), pp. 113–125, 2002.

[2] Lampart, P. & Hirt Ł., Complex multidisciplinary optimization of turbine blading systems. *Archives of Mechanics*, **64**(2), pp. 153–175, 2012.

[3] Al Jubori, A., Al-Dadah, R.K., Mahmoud, S., Bahr Ennil, A.S. & Rahbar, K., Three dimensional optimization of small-scale axial turbine for low temperature heat source driven organic rankine cycle. *Energy Conversion and Management,* **133**, pp. 411–426, 2016.
https://doi.org/10.1016/j.enconman.2016.10.060

[4] Elmegaard, B. & Brix, W., Efficiency of Compressed Air Energy Storage, *24th International Conference on Efficiency, Cost, Optimization, Simulation and Environmental Impact of Energy Systems*, pp. 1–12, 2012.

[5] Klonowicz, P., Heberle, F., Preißinger, M. & Brüggemann, D., Significance of loss correlations in performance prediction of small scale, highly loaded turbine stages working in organic rankine cycles. *Energy*, **72**, pp. 322–330, 2012.
https://doi.org/10.1016/j.energy.2014.05.040

[6] Beyene, A. & Miller, A., Krytyczna ocena metod określania strat w turbinowych palisadach profile. *Biuletyn Informacyjny Instytutu Techniki Cieplnej Politechniki Warszawskiej*, pp. 19–34, 1986. (in Polish).

[7] Kosowski, K., Tucki, K. & Kosowski, A., Turbine stage design aided by artificial intelligence methods. *Expert Systems with Applications,* **36**(9), pp. 11536–11542, 2009. https://doi.org/10.1016/j.eswa.2009.03.053

[8] Pierret, S. & Van den Braembussche, R.A., Turbomachinery blade design using a Navier-Stokes solver and artificial neural network. *Journal Turbomachinery*, **121**, pp. 326–332, 1999. https://doi.org/10.1115/1.2841318

[9] Forrester, A.I.J., Sóbester, A. & Keane, A.J., *Engineering Design via Surrogate Models*, Wiley, pp. 23–24, 2008.

[10] Hedar, A. & Fukushima, M., Minimizing multimodal functions by simplex coding. *Optimization Methods and Software*, **18**(3), pp. 265–282, 2003. https://doi.org/10.1080/1055678031000098773a

[11] Rodriguez, F.J., García-Martinez, C. & Lozano, M., Hybrid metaheuristics based on evolutionary algorithms and simulated annealing: Taxonomy, comparison, and synergy test. *IEEE Transactions on Evolutionary Computation*, **16**, pp. 787–800, 2012. https://doi.org/10.1109/tevc.2012.2182773

[12] Mahmuddin, M. & Yousof, Y., A hybrid simplex search and bio-inspired algorithm for faster convergence. *International Conference Machine Learning and Computing*, **3**, pp. 203–207, 2011.

[13] Mirsadeghi, E. & Shariat Panahi, M., Hybridizing artificial bee colony with simulated annealing. *International Journal Hybrid Information Technology,* **5**, pp. 11–18, 2012.

[14] Jovanovic, R., Kais, S. & Alharbi, F.H., Cuckoo search inspired hybridization of the Nelder-Mead simplex algorithm applied to optimization of photovoltaic cells. *Applied Mathematics Information Science,* **10**, pp. 961–973, 2016. https://doi.org/10.18576/amis/100314

[15] Yang, X.Y., A New Metaheuristic Bat-Inspired Algorithm. In: Nature Inspired Cooperative Strategies for Optimization (NISCO 2010), *Springer Berlin Heidelberg*, pp. 65–74, 2010.

[16] Nelder, J.A. & Mead, R., A simplex method for function minimization. *Computer Journal*, **7**(4), pp. 308–313, 1965. https://doi.org/10.1093/comjnl/7.4.308

[17] Yilmaz, S. & Küçüksille, E.U., A new modification approach on bat algorithm for solving optimization problems. *Applied Soft Computing Journal,* **28,** 259–275, 2015. https://doi.org/10.1016/j.asoc.2014.11.029

[18] Rezaee Jordehi, A., Chaotic bat swarm optimisation (CBSO). *Applied Soft Computing Journal*, **26**, pp. 523–530, 2015. https://doi.org/10.1016/j.asoc.2014.10.010

[19] Yang, X. & Gandomi, A.H., Bat algorithm: a novel approach for global engineering optimization. *Engineering Computations*, **29**(5), pp. 464–483, 2012. https://doi.org/10.1108/02644401211235834

[20] Spendley, W., Hext, G. & Himsworth, F., Sequential application of simplex designs in optimisation and evolutionary operation. *Technometrics*, **4**(4), pp. 441–461, 1962. https://doi.org/10.1080/00401706.1962.10490033

[21] Yang, X.S., *Recent Advances in Swarm Intelligence and Evolutionary Computation*, Springer, 2015.
[22] Lampart, P., Numerical Optimization of Stator Blade Sweep and Lean in an LP Turbine Stage. *International Joint Power Generation Conference, ASME*, pp. 579–592, 2002.
[23] Lampart, P., Control of LP turbine rotor blade underloading using stator blade compound lean at root. *Journal of Thermal Science*, **9**(2), pp. 115–121, 2000. https://doi.org/10.1007/s11630-000-0004-3
[24] Rosic, B. & Xu, L., Blade lean and shroud leakage flows in low aspect ratio turbines. *Journal of Turbomachinery*, **134**(3), pp. 31003-1–31003-12, 2012. https://doi.org/10.1115/1.3106002
[25] Klonowicz, P., Surwiło, J., Witanowski, Ł., Suchocki, T., Kozanecki, Z. & Lampart, P., Design and numerical study of turbines operating with MDM as working fluid. *Open Engineering*, **5**, pp. 485–499, 2015. https://doi.org/10.1515/eng-2015-0050
[26] Witanowski, Ł., Klonowicz, P. & Lampart, P., UWCAES Optimization of a radial-axial turbine working in a UWCAES system. *Mechanik*, **7**, pp. 3–9, 2016.
[27] Lagarias, J., Reeds, J., Wright, M. & Wrightm P., Convergence properties of Nelder-Mead simplex method in low dimensions. *SIAM Journal on Optimization*, **9**, pp. 112–147, 1998. https://doi.org/10.1137/s1052623496303470
[28] Kelly, C.T., *Implicit Filtering*, North Carolina State University, Raleigh, North Carolina, 2011.

ENERGY AND EXERGY EFFICIENCIES OF DIFFERENT CONFIGURATIONS OF THE EJECTOR-BASED CO_2 REFRIGERATION SYSTEMS

S. TASLIMITALEGHANI, M. SORIN & S. PONCET
Department of Mechanical Engineering, Université de Sherbrooke, Sherbrooke (QC), Canada.

ABSTRACT

Carbon dioxide (CO_2) is an appropriate replacement for conventional refrigerants due to its low global warming effects. However, its application within a traditional refrigeration compression cycle leads to low thermodynamic performance due to the large expansion losses in a throttling process. The application of ejectors allows reducing these losses. Many scenarios of ejector-based cycles have been proposed. Among them four different configurations may be distinguished: an expansion work recovery cycle (EERC), a liquid recirculation cycle (LRC), an increasing compressor discharge pressure cycle (CDPC) and a vapor jet refrigeration cycle (VJRC). This study deals with the comparative analysis of these cycles. In order to study the performance of the cycles, the numerical simulations are developed using EES software. Two performance criteria, energy efficiency (COP) and exergy efficiency are evaluated for each cycle. The highest values of these criteria point to the most thermodynamically efficient cycle. The results show that the EERC has the highest COP and exergy efficiency compared to other cycles. For example, the COP of the EERC is 3.618 and the exergy efficiency is 9.68%. The COP (resp. exergy efficiency) is approximately 23.3% (resp. 23.3%), 24.9% (resp. 25.5%) and 5.6 times (resp. 56.2%) higher than the corresponding energy and exergy efficiencies of LRC, CDPC and VJRC. Moreover, in comparison with a basic throttling valve cycle, the COP and exergy efficiency in EERC are higher up to 23% and 24% correspondingly. The detailed exergy analysis of EERC cycle has pinpointed the equipment where the major exergy losses take place. The largest losses occur in the evaporator (about 33% of the total exergy destruction of the cycle) followed by the compressor (25.5%) and the ejector (24.4%).

Keywords: comparative analysis, COP, ejector, exergy efficiency, refrigeration systems, transcritical CO_2 cycles, two-phase.

1 INTRODUCTION

Carbon dioxide (CO_2) is an appropriate replacement for conventional refrigerants due to its low global warming effects. One of the disadvantage of the cycle is a large exergy loss due to an important pressure reduction during expansion of CO_2 from the supercritical to the sub-critical state in a throttling valve. Among different devices for expansion work recovery, ejector is a favourable equipment, which enables to reduce losses by recovering part of the expansion work in a throttling process and improve the cycle's efficiency.

The first application of two-phase ejector to the transcritical CO_2 cycle was first described by Gay [1]. It was proposed to replace the expansion valve by a two-phase ejector to reduce the losses due to the throttling process.

Kornhauser [2] was the first to develop a one-dimensional and homogeneous model of a two-phase ejector using R12 as a refrigerant in the ejector expansion refrigeration system (EERS).

Li and Groll [3] adapted the Kornhauser's model for an ejector used within a transcritical CO_2 air-conditioning system.

A thermodynamic exergy analysis of transcritical CO_2 ejector refrigeration system was performed by Fangtian and Yitai [4]. They evaluated COP and exergy destruction of the

DOI: 10.2495/EQ-V3-N1-22-33

system. Their results showed an improvement of 30% in COP and decreasing exergy loss more than 25% compared to the conventional system.

The present study is focused on a thermodynamic comparative analysis of the performance of the different transcritical CO_2 ejector cycles to identify the most efficient one. The COP, exergy efficiency and exergy destructions are calculated and compared for the expansion work recovery cycle (EERC), liquid recirculation cycle (LRC), increasing compressor discharge pressure cycle (CDPC) and vapor jet refrigeration cycle (VJRC). Exergy analysis is employed to determine the amount and locations of irreversibilities within different components of each cycle.

2 EJECTOR APPLICATIONS FOR TRANSCRITICAL CO_2 CYCLES

Different applications of the ejector in CO_2 air-conditioning and refrigeration systems used in this study are as follows:

- Ejector for utilization of low-grade energy (vapor jet ejector systems, VJRC)
- Ejectors for expansion work recovery cycle (standard two-phase ejector, EERC)
- Ejectors for liquid recirculation cycle (LRC)
- Ejector for increasing compressor discharge pressure (CDPC)

2.1 Vapor jet ejector systems (single-phase ejectors)

In the vapor jet cycle, a pump, a generator, and an ejector replace the compressor. A fraction of the liquid from the condenser is pumped to a high pressure and temperature. The fluid absorbs heat at a constant pressure from a low-grade energy source in the generator. The heated flow expands in a primary nozzle to a high velocity and a low pressure. This low pressure entrains the secondary flow from the evaporator into the mixing chamber of the ejector. The irreversible mixing of the two fluids occurs in the mixing chamber depending on the ejector geometry at the constant pressure or at the constant area. Finally, the flow decelerates in the diffuser by converting the remaining kinetic energy into the pressure increase. The vapor exiting the diffuser is condensed at a constant pressure. The liquid at the condenser exit is pumped to the generator. The vapor is sent through the metering valve to the evaporator.

The main advantage of the VJRCs is that they can produce a refrigeration effect by using the low-grade waste heat for heating the primary flow in the generator.

Compared to a conventional system, for the same pressure increase, the work of the liquid pump in the VJRC is less than the compressor work and it does not also require any lubrication [5]. The schematic of a transcritical CO_2 VJRC and corresponding temperature-specific entropy diagram are shown in Fig. 1. It can be seen the flow through the mixing section and the diffuser remains vapor so the ejector works in a single-phase mode.

2.2 Two-phase ejectors for expansion work recovery (EERC)

A two-phase ejector can be used in vapor compression systems for recovery of the expansion work by reducing the throttling losses to improve the performance of the system.

As shown in Fig. 2, the subcritical CO_2 coming from the vapor port of the separator is compressed to high pressure and temperature to the supercritical state. It releases heat in the gas cooler. After the gas cooler exit, the stream enters the primary nozzle of the ejector and expands at the mixing section. The secondary vapor stream pre-accelerates into the mixing section. The

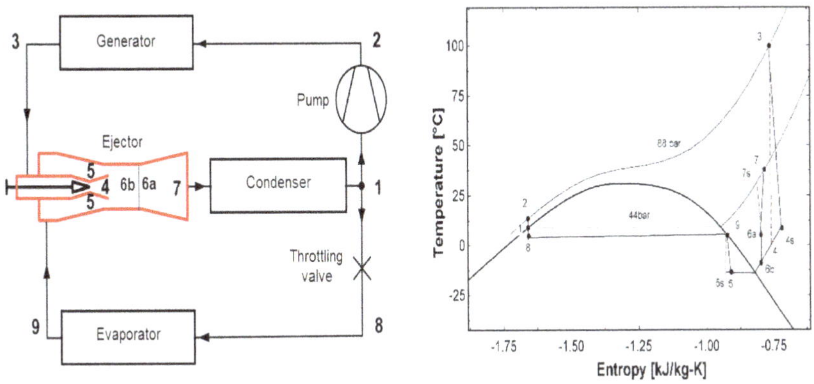

Figure 1: Transcritical CO_2 vapor jet refrigeration cycle and corresponding temperature-specific entropy diagram.

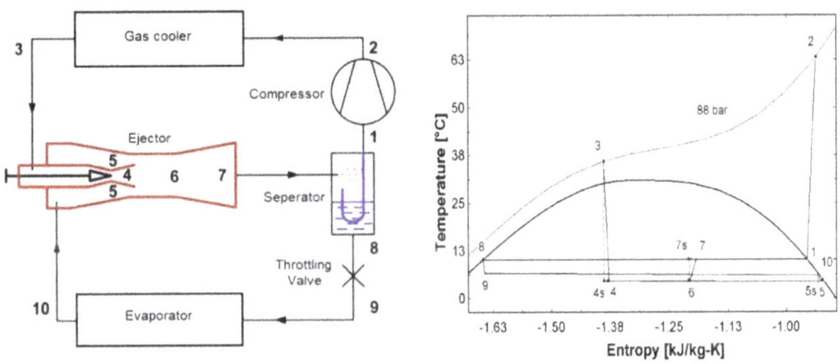

Figure 2: Transcritical CO_2 ejector expansion recovery cycle and the corresponding temperature-specific entropy diagram.

mixture then flows through the diffuser what causes a compression before entering the separator. Vapor portion of the two-phase flow returns to the compressor while the pressure of the liquid portion is reduced through the metering valve before entering the evaporator. The stream absorbs heat in the evaporator before it enters the ejector.

EERC has two main advantages. First, the cooling capacity increases because the isentropic expansion inside the primary nozzle in comparison to an isenthalpic expansion valve of a conventional system has a larger enthalpy difference. Second, the compressor work is decreased due to the increase of the suction pressure of the compressor resulting in COP improvement.

2.3 Two-phase ejectors for liquid recirculation

In this cycle, the ejector is used to recirculate liquid and improve the evaporator performance. It was first patented by Phillips [6] and later by Lorentzen [7]. Figure 3 shows a schematic of the cycle and corresponding T-S diagram for transcritical CO_2. The expansion work

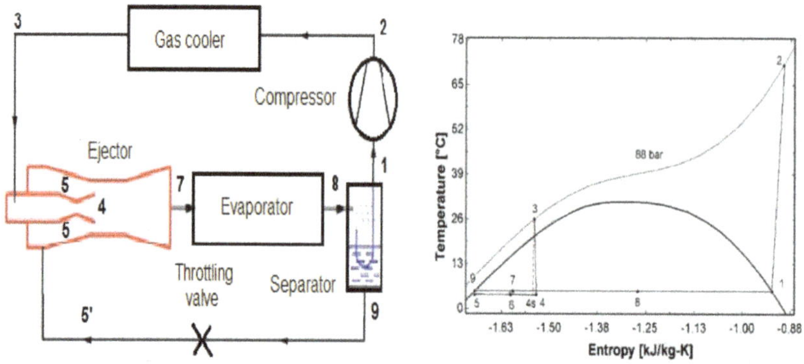

Figure 3: Transcritical CO_2 liquid recirculation cycle and the corresponding temperature-specific entropy diagram.

recovered by the ejector is used to liquid recirculation. The amount of COP improvement by using liquid recirculation is dependent on the working fluid used. There is also an optimum value for every system and operating condition because with increasing the recirculation both the heat transfer coefficient and the pressure drop increase [8].

Lawrence and Elbel [8] studied two applications of the two-phase ejector cycle for CO_2 as a working fluid: first was liquid recirculation cycle that used the ejector to improve the evaporator performance and other was a standard two-phase ejector that used work recovery of the ejector to increase the compressor pressure. The COP improvement of 3% through CO_2 ejector was obtained in the recirculation cycle as it could reach up to 25% in a standard two-phase ejector to directly unload the compressor pressure.

2.4 Two-phase ejector for increasing compressor discharge pressure

This innovative cycle was recently introduced by Bergander [9]. In this cycle, ejector is used as a second-stage compressor. Unlike standard two-phase ejector which increases the suction pressure of the compressor, in this cycle, the ejector is used to increase the compressor discharge pressure. In a subsequent work, Bergander [10] developed a thermodynamic ejector model for R22 and conducted experiments that showed 16% COP improvement. In this cycle, there is a two-phase flow inside the ejector, liquid for the primary flow and vapor for the secondary one. The primary flow enters the ejector after exiting the pump and mixes with the secondary flow that comes from the compressor. The flow at the exit of the ejector enters the gas cooler. The layout of this cycle and corresponding T-S diagram are presented in Fig. 4.

3 EXERGY ANALYSIS OF DIFFERENT EJECTOR CYCLES

The exergy analysis of the ejector cycles introduced in Section 2 is carried out to investigate the exergy destruction of the different components of the system to determine the maximum performance and potential improvements of the cycles.

3.1 Modeling of two-phase flow ejector

Different models of the ejectors exist according to assumptions, governing equations, auxiliary conditions, mixing mechanism and solution methods. Thermodynamic modeling is a simple way to solve the equations in one dimension. It is also easily integrated into a system.

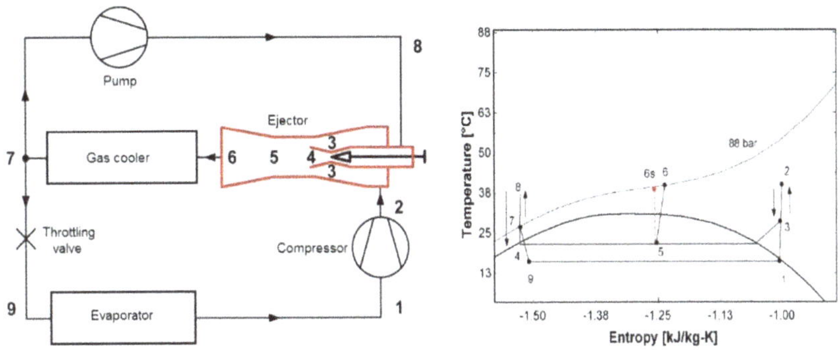

Figure 4: CO_2 transcritical ejector cycle to increase the compressor discharge pressure and corresponding temperature-specific entropy diagram.

Conservation equations of mass, energy and momentum, some gas dynamic equations, state equations, isentropic relations as well as some appropriate assumptions, initial and boundary conditions are used to solve the flow within the ejector. Some assumptions that are usually employed to simplify the problem are as follows: adiabatic walls of the ejector, steady-state flow, isentropic efficiencies for the nozzles and the diffuser, stagnation points of the streams at inlets and outlet of the ejector and mixing coefficient for mixing losses.

Most ejector models presented for CO_2 two-phase flows are based on a homogeneous equilibrium model in which both gas and liquid are in thermodynamic and mechanical equilibrium. It means that both phases have the same pressure, temperature, velocity, turbulence kinetic energy and turbulence dissipation rate [2, 3, 11–16].

3.1.1 Assumptions and calculation procedure

The thermodynamic model of the two-phase ejectors (EERC, LRC, CDPC, section 2.2~2.4) is based on the following assumptions:

1. Flow is one-dimensional, steady state and adiabatic through the ejector.
2. The homogeneous equilibrium is assumed for two-phase flow.
3. The CO_2 thermodynamic and transport properties of the primary and secondary flows are obtained from real fluids properties.
4. Flow losses in the pipes and heat exchangers are negligible.
5. Kinetic energies of the refrigerant are negligible at the ejector inlet and outlet.
6. Friction losses are defined in terms of isentropic efficiencies in the nozzles, diffuser and mixing.
7. Mixing occurs under a constant pressure in the ejector mixing section with the assumption that the fluid momentum is conserved.
8. Pressure loss of the secondary flow is assumed $\Delta P = 1bar$ for EERC, LRC, CDPC.
9. Critical-mode operation is applied for VJRC and normal shock takes place at the end of the constant area mixing chamber [17, 18] .
10. The secondary inlet flow is considered as a saturated vapor in EERC, saturated liquid in LRC, superheated vapor in CDPC.
11. The heat sink temperature (or the ambient temperature) is 35°C for EERC, LRC, CDPC and $\left(T_{gen,in} + T_{gen,out}\right)/2 + 5°C$ for VJRC; the heat source temperature is 27°C.

Table 1: Constant parameters used in the simulation.

Pressure of gas cooler	88 bar	Cooling capacity	72 kW
Temperature of gas cooler exit	36°C	Efficiency of Nozzles	0.8
Temperature of evaporator	5°C	Efficiency of diffuser	0.8
Temperature of generator exit (VJRC)	100°C	Efficiency of mixing	0.95
Pressure of generator (VJRC)	88 bar	Efficiency of compressor	0.75
Pressure of evaporator	39.69 bar	Efficiency of pump	0.75

The constant parameters used in the simulations of the cycles are shown in Table 1.

An engineering equation solver (EES) program is used to solve the proposed models in Section 2, which combines non-linear equations with thermo-physical property functions.

The modeling of the ejector expansion recovery cycle is based on one unit of mixing refrigerant mass flow in the mixing sector of the ejector. Therefore, the primary mass flow from the gas cooler is $1/(1+\mathrm{ER})$ and the secondary mass flow from the evaporator is ER/(1 + ER).

The model is solved according to the relationship between the vapor quality of the ejector outlet and the entrainment ratio. The solution converges when eqn (1) is satisfied to maintain a balance between liquid and vapor in the expansion recovery cycle:

$$x_{out,diff} = \frac{1}{1+ER} \tag{1}$$

First the properties at different states of the cycles are calculated. In EERC, according to Fig. 3, the specific enthalpy at gas cooler and evaporator exit (h_3, h_{10}) are defined. The motive flow expands to mixing pressure with a nozzle efficiency η_{pn} defined as:

$$\eta_{pn} = \frac{h_3 - h_4}{h_3 - h_{4s}} \tag{2}$$

where

$$\mathrm{h}_{4s} = \mathrm{f}\left(\mathrm{P}_{evap} - \Delta P, \mathrm{s}_3\right) \tag{3}$$

By applying energy conservation law between state 3 and state 4, the velocity at state 4 is obtained:

$$\frac{1}{2}u_4^2 = h_3 - h_4 \tag{4}$$

The velocity of secondary flow (u_5) is calculated in the same way as that of the primary flow and then the velocity of mixed flow is determined by the momentum equation in mixing chamber according to:

$$u_6\,\mathrm{m}_6 = \mathrm{m}_4 u_4 + \mathrm{m}_5 u_5 \tag{5}$$

The mixing efficiency is defined as [19] :

$$\eta_{mix} = \frac{\frac{1}{2}m_{4'}u_{4'}^2}{\frac{1}{2}m_4 u_4^2} \tag{6}$$

where $u_{4'}$ is the corrected velocity at state 4 which takes into account the mixing loss.
The energy conservation between two inlets and outlet of the ejector is as follows:

$$m_3 h_3 + m_{10} h_{10} = m_7 h_7 \tag{7}$$

The energy conservation between inlet and outlet of the diffuser is described as:

$$\frac{1}{2} u_6^2 + h_6 = h_5 \tag{8}$$

The diffuser efficiency is defined as:

$$\eta_{diff} = \frac{h_{7s} - h_6}{h_7 - h_6} \tag{9}$$

The pressure and quality at the ejector outlet (state 7) are obtained as

$$p_7 = f\left(s_6, h_{7s}\right) \tag{10}$$

$$x_7 = f\left(p_7, h_7\right) \tag{11}$$

This quality satisfies the eqn (1). The cooling capacity of the cycles (Figs. 1–4) can be written:

$$Q_{ev} = m_{evap}\left(h_{out,evap} - h_{in,evap}\right) = \frac{ER}{\left(1+ER\right)}\left(h_{out,evap} - h_{in,evap}\right) \tag{12}$$

The compressor power consumption is

$$W_{comp} = m_{comp}\left(h_{out,comp} - h_{in,comp}\right) = \frac{1}{\left(1+ER\right)}\left(h_{out,comp} - h_{in,comp}\right) \tag{13}$$

The gas cooler capacity is:

$$Q_{gc} = m_{gc}\left(h_{out,gc} - h_{in,gc}\right) = \frac{1}{\left(1+ER\right)}\left(h_{out,gc} - h_{in,gc}\right) \tag{14}$$

The cooling coefficient of performance (COP_c)for EERC and LRC is obtained using:

$$COP_C = \frac{Q_{evap}}{W_{comp}} \tag{15}$$

For the cycle including the pump (CDPC) the COP_c is defined as

$$COP_C = \frac{Q_{evap}}{W_{comp} + W_{pump}} \tag{16}$$

For vapor jet refrigeration system, VJRC, it is expressed as

$$COP_C = \frac{Q_{evap}}{Q_{gen} + W_{pump}} \tag{17}$$

3.1.2 Exergy calculations

The exergy in all states is calculated based on the unit mass flow of mixing refrigerant in the ejector:

$$ex_k = m_k \cdot \left[\left(h_k \quad h_0 \right) - T_0 \cdot \left(s_k \quad s_0 \right) \right] \tag{18}$$

where m_k is the mass flow at the cycle state k.

The exergy destructions in the various processes are calculated as follows:

Compressor:

$$ex_{loss,comp} = ex_{in,comp} - ex_{out,comp} + W_{comp} \tag{19}$$

Gas cooler:

$$ex_{loss,gc} = ex_{in,gc} - ex_{out,gc} + Q_{gc} \cdot \left[1 - \left(\frac{T_0}{T_{sink}} \right) \right] \tag{20}$$

Ejector:

$$ex_{loss,ej} = ex_{in,pn} + ex_{in,sn} - ex_{out,diff} \tag{21}$$

Evaporator:

$$ex_{loss,evap} = ex_{in,evap} - ex_{out,evap} + Q_{evap} \cdot \left[1 - \left(\frac{T_0}{T_{source}} \right) \right] \tag{22}$$

Throttling valve:

$$ex_{loss,th} = ex_{in,th} - ex_{out,th} \tag{23}$$

The total exergy destruction is calculated using:

$$ex_{loss,tot} = ex_{comp} + ex_{ej} + ex_{th} + ex_{evap} + ex_{gc} + ex_{gen}\left(VJRC\right) + ex_{cond}\left(VJRC\right) \tag{24}$$

The exergy efficiency of the cycles is evaluated as:

$$\eta_{ex} = 1 - \frac{ex_{loss,tot}}{W_{comp}} \tag{25}$$

For the cycle includes the pump, CDPC, exergy efficiency is as following:

$$\eta_{ex} = 1 - \frac{ex_{loss,tot}}{W_{comp} + W_{pump}} \tag{26}$$

For VJRC (Fig. 1), exergy efficiency and total exergy loss are calculated by:

$$\eta_{ex} = 1 - \frac{ex_{loss,tot}}{W_{pump} + ex_{Q,ge} + ex_{Q,cond}} \tag{27}$$

$$ex_{loss,tot} = ex_{pump} + ex_{gen} + ex_{cond} + ex_{ej} + ex_{th} + ex_{ev} \tag{28}$$

where $ex_{Q,gen}$ and $ex_{Q,cond}$ are exergy transfer by heat in generator and condenser respectively which are defined as:

$$ex_Q = Q . \left[1 - \left(\frac{T_0}{T} \right) \right] \tag{29}$$

4 RESULTS

The comparison of the results for four transcritical CO_2 refrigeration cycles is presented below. Exergy destructions and exergy efficiencies of the cycles are calculated under constant cooling capacity and corresponding parameters listed in Table 1.

As shown in Tables 2 and 3, the EERC has the highest COP and exergy efficiency. For the high-side pressure of 88 bar, the COP for EERC is 23.3%, 24.9% and 5.6 times higher than the COP for LRC, CDPC and VJRC, respectively.

It is also shown that EERC improves the COP by up to 23.1% compared to basic cycle without the ejector, while the COP of LRC and CDPC remains almost constant and for VJRC, the COP is very low. For given operating conditions, the pressure ratio of EERC is also the largest among other cycles.

Table 2: Comparison of the ejector's performance of the cycles.

Device	Ejector performance				
	EERC	LRC	CDPC	VJRC	BC
COP	3.618	2.935	2.896	0.6476	2.938
ER	0.564	0.641	1.558	0.921	-
P_{ratio}	1.15	1.03	1.01	1.11	-

Table 3: Exergy destructions and exergy efficiencies of the cycles (P_{gc} = 88 bar, T_{evap} = 5°C, Q_{evap}=72KW).

Device	Exergy Loss, Kw									
	EERC		LRC		CDPC		VJRC		BC	
	Loss, kW	%	Loss, kW	%	Loss, kW	%	Loss, kW	%	Loss, kW	%
Compressor	4.58	25.5	5.539	24.55	5.475	23.86	–	–	5.533	24.5
Gas cooler	2.817	15.68	3.515	15.55	1.679	7.317	–	–	3.511	15.55
Ejector	4.382	24.38	7.705	34.08	2.082	9.074	12.82	43.91	–	–
Valve	0.338	1.878	–	–	7.696	33.54	0.2198	0.753	7.696	34.07
Evaporator	5.854	32.57	5.847	25.86	5.847	25.48	5.808	19.89	5.847	25.88
Generator	–	–	–	–	–	–	4.931	16.89	–	–
Condenser	–	–	–	–	–	–	4.73	16.2	22.59	–
Pump	–	–	–	–	0.165	0.72	0.686	2.35	–	–
Total	17.97	100	22.61	100	22.94	100	29.2	100	24.51	100
W_{comp}	19.9	–	24.53	–	24.19	–	–	–	–	–
W_{pump}	–	–	–	–	0.668	–	2.558	–	–	–
$ex_{Q,evap}$	-1.919	–	-1.919	–	–	–	-1.919	–	-1.919	–
$ex_{Q,,gen}$	–	–	–	–	–	–	8.8	–	–	–
$ex_{Q,cond}$	–	–	–	–	–	–	19.76	–	–	–
Q_{gen}	–	–	–	–	–	–	108.6	–	–	–
η_{ex}	–	9.683	–	7.856	–	7.718	–	6.197	–	7.831

Table 3 shows the exergy losses in each component and exergy efficiency of all cycles. It can be noticed that EERC has the maximum exergy efficiency. The entrainment ratio and pressure ratio are 0.564 and 1.15 for EERC, respectively. The exergy loss in the evaporation process is the largest one in this system, while for LRC and VJRC, the largest loss occurs in the ejector.

The throttling exergy loss in the basic cycle is 7.69 KJ kg^{-1} that constitutes 34.07% of the total exergy loss. However, it is only 0.34 KJ kg^{-1}, 1.88% of the total exergy loss in EERC and the ejector's exergy loss is also 4.383 KJ kg^{-1}, 24.4% .The sum of these two losses is 26.28% of the total exergy loss of the system which is less than the throttling loss in the conventional cycle. The exergy loss in compressor and gas cooler are also reduced in EERC, and it is almost constant in the evaporator.

The use of liquid recirculation in refrigeration system improves the entrainment ratio compared to EERC; however, COP and exergy efficiency remain constant compared to conventional cycle. Therefore, despite a large amount of work that can be recovered with the CO_2 ejector, there is not the COP improvement for LRC.

The CDPC simulation shows that the ejector integration with this cycle is not efficient. It is due to the fact that the pressure lift is accomplished mainly by the compressor not the ejector. The pressor ratio is obtained 1.01 for this cycle.

The jet refrigeration cycle (single phase ejector, Fig. 1) achieves the lowest COP and the lowest exergy efficiency compared to other cycles. The low COP value around 0.65 is obtained. The high exergy losses of heat transfer in condenser (16.2%) and generator (16.9%) result in low exergy efficiency.

5 CONCLUSION

A comparative study based on the first and second laws of thermodynamics is performed for different transcritical CO_2 refrigeration cycles that use an ejector: EERC, LRC, CDPC and VJRC. The analysis for given conditions led to the following conclusions:

- Transcritical CO_2 refrigeration cycles, EERC has the highest COP and exergy efficiency. For the given operating conditions, it improves the COP and exergy efficiency by up to 23% and 24%, respectively, compared to the basic throttling cycle.
- In EERC, the irreversibility loss of the expansion process is significantly reduced compared to basic throttling valve cycle and as a result the exergy efficiency is increased.
- The COP of EERC is improved by up to 23.3%, 24.9% and 5.6 times compared to the LRC, CDPC and VJRC.
- The exergy loss in the evaporation process is the largest loss in EERC, whereas for LRC and VJRC the ejection process has the largest loss.
- The use of liquid recirculation improves entrainment ratio compared to EERC, however, COP and exergy efficiency decrease.
- CO_2 can gain more benefit from EERC compared to other cycles. CO_2 ejector liquid recirculation cycle and VJRC has a low potential for COP improvement.
- Ejector is not effective in the cycle for increasing compressor discharge pressure because the pressure lift is mainly accomplished by the compressor.

Nomenclature

Symbol/ Abbreviation		Greek	
BC	Basic cycle without ejector	η	isentropic efficiency
CDPC	Increasing compressor discharge pressure cycle	ρ	density, kg/m^3
COP	Coefficient of performance	**subscripts**	
EERC	Expansion work recovery cycle	0	dead state in the exergy analysis
ER	Entrainment ratio	C	cooling
ex	Exergy	comp	compressor
Ex_Q	Exergy transferred by heat	cond	condenser
h	Specific enthalpy, kJ/kg	diff	diffuser
LRC	Liquid recirculation cycle	ej	ejector
m	Mass flow rate, kg/s	evap	evaporator
P	Pressure, kPa	gc	gas cooler
P_{ratio}	Pressure ratio (pressure lift)	gen	Generator
Q	Heat rate, kW	in	Inlet
s	Specific entropy, kJ/kg K	mix	Mixing
T	Temperature, K	out	Outlet
u	Velocity, m/s	pn	primary nozzle
VJRC	Vapor jet refrigeration cycle	sn	secondary nozzle
W	Work rate, kW	th	Throttling
x	Quality	tot	Total
ΔP	Pressure drop		
η_{ex}	Exergy efficiency		

ACKNOWLEDGEMENTS

This project is a part of the Collaborative Research and Development (CRD) Grants Program at 'Université de Sherbrooke'. The authors acknowledge the support of the Natural Sciences and Engineering Research Council of Canada, Hydro-Québec, Rio Tinto, Alcan and Canmet ENERGY Research Center of Natural Resources Canada.

REFERENCES

[1] Gay, N.H., Refrigerating system. Google Patents, 1931.

[2] Kornhauser, A.A., The use of an ejector as a refrigerant expander. *Proceedings of the 1990 USNC/IIR – Purdue Refrigeration Conference*, Purdue University, West Lafayette, IN, US, pp. 10–19, 1990.

[3] Li, D. & Groll, E.A., Transcritical CO_2 refrigeration cycle with ejector-expansion device. *International Journal Refrigeration*, **28**(5), pp. 766–773, 2005. https://doi.org/10.1016/j.ijrefrig.2004.10.008

[4] Fangtian, S. & Yitai, M., Thermodynamic analysis of transcritical CO_2 refrigeration cycle with an ejector. *Applied Thermal Enginnering,* **31**(6), pp. 1184–1189, 2011. https://doi.org/10.1016/j.applthermaleng.2010.12.018

[5] Elbel, S. & Hrnjak, P., Ejector refrigeration: an overview of historical and present developments with an emphasis on air-conditioning applications. *International Refrigeration and Air Conditioning Conference*, pp. 2350-1–2350-8, 2008.

[6] Phillips, H.A., Refrigeration system. Google Patents, 1938.

[7] Lorentzen, G., Throttling, the internal haemorrhage of the refrigeration process. *Proceedings of the Institute of Refrigeration*, **80**, pp. 39–47, 1983.

[8] Lawrence, N. & Elbel, S., Mathematical modeling and thermodynamic investigation of the use of two-phase ejectors for work recovery and liquid recirculation in refrigeration cycles. *International Journal Refrigeration*, **58**, pp. 41–52, 2015. https://doi.org/10.1016/j.ijrefrig.2015.06.004

[9] Bergander, M.J, *New regenerative cycle for vapor compression refrigeration, Final Scientific Report,* DOE Award DE-FG36-04GO14327. Madison, CT, USA, 2005.

[10] Bergander, M., Refrigeration cycle with two-phase condensing ejector, *International Refrigeration and Air Conditioning Conference*, Purdue, USA, pp. R008-1-R008-8, 2006.

[11] Elbel, S.W. & Hrnjak, P.S., Effect of internal heat exchanger on performance of transcritical CO_2 systems with ejector. *International Refrigeration and Air Conditioning Conference*, pp. R166-1–R166-8, 2004.

[12] Li, D., *Investigation of an ejector-expansion device in a transcritical carbon dioxide cycle for military ECU applications* [Ph. D. Dissertation]. Purdue University, West Lafayette, IN, 2006.

[13] Nehdi, E., Kairouani, L. & Bouzaina, M., Performance analysis of the vapour compression cycle using ejector as an expander. *International Journal Energy Research*, **31**(4), pp. 364–375, 2007. https://doi.org/10.1002/er.1260

[14] Sarkar, J., Optimization of ejector-expansion transcritical CO_2 heat pump cycle. *Energy,* **33**(9), pp. 1399–1406, 2008. https://doi.org/10.1016/j.energy.2008.04.007

[15] Ameur, K., Aidoun, Z. & Ouzzane, M., Modeling and numerical approach for the design and operation of two-phase ejectors. *Applied Thermal Engineering,* **109**, pp. 809–818, 2016. https://doi.org/10.1016/j.applthermaleng.2014.11.022

[16] Smolka, J., Bulinski, Z., Fic, F., Nowak, A.J., Banasiak, K. & Hafner, A., A computational model of a transcritical R744 ejector based on a homogeneous real fluid approach. *Applied Mathematical Modelling,* **37**(3), pp. 1208–1224, 2013. https://doi.org/10.1016/j.apm.2012.03.044

[17] Khalil, A., Fatouh, M. & Elgendy, E., Ejector design and theoretical study of R134a ejector refrigeration cycle. *International Journal Refrigeration*, **34**(7), pp. 1684–1698, 2011. https://doi.org/10.1016/j.apm.2012.03.044

[18] Selvaraju, A. & Mani, A., Analysis of an ejector with environment friendly refrigerants. *Applied Thermal Engineering*, **24**(5–6), pp. 827–838, 2004. https://doi.org/10.1016/j.applthermaleng.2003.08.016

[19] He, Y., Deng, J. & Zhang, Z., Thermodynamic study on a new transcritical CO_2 ejector expansion refrigeration system with two-stage evaporation and vapor feedback. *HVACR Research*, **20**(6), pp. 655–664, 2014. https://doi.org/10.1080/10789669.2014.929422

PYROLYSIS BEHAVIORS OF WASTE COCONUT SHELL AND HUSK BIOMASSES

Q. WANG & J. SARKAR
Graduate School of Science and Engineering Saitama University, Japan.

ABSTRACT

Coconut shell and husk are two biomasses wastes abundant in most of the coastal countries. However, despite their enormous potential as energy sources, they are hardly studied and their thermal characteristics are still not well known. In this study, both biomasses are thermally degraded through thermogravimetry (TG-DTA) and their pyrolysis product yield such as char, tar and gases are analyzed. The TG-DTA results show that pyrolysis of biomass consists of three stages. Three stages can be outlined as: (1) dehydration process for temperatures below 122°C, (2) pyrolytic cracking from 122°C to 400°C, stage consist of two exothermic simultaneous processes where hemicelluloses, cellulose and lignin are decomposed and a high amount of volatile matter formation occurs and (3) the last endothermic decomposition of the lignin at temperatures above 400°C. From the pyrolytic results, it is showed that the char and gases yields were increased with the decrement tar. The gas-evolving profiles from pyrolyzing the coconut shell and husk components in a packed bed, monitored by a GC-TCD and a GC-FID, showed similar behavior. H_2 was released out at a higher temperature (>450°C) and it got the maximum rate at 700°C then it decreased. CO_2 was released out at 130°C–750°C and got the maximum releasing value at 300°C–400°C. The released CO showed almost similar pattern with that of CO_2. However, the release rate was lower than CO_2 and the maximum release rate of CO was found at 300°C –400°C. CH_4 was released out at the temperature between 200°C –850°C, and it got the maximum rate at 550°C. The releasing of hydrocarbon was generally very low.

Keywords: biomass, fixed bed, pyrolysis, TGA, thermogravimetry.

1 INTRODUCTION

Modern energy enriches life. There are 7 billion people on earth who use energy each day to make their lives richer, more productive, safer and healthier. Energy is the driving force for the development of world economy. Due to the ever-increasing world population and the continually improving living standards, the demand for energy resources has risen dramatically, and this condition will continue. Fossil fuels such as petroleum, coal and natural gas satisfy the major fraction of total need of world's energy. However, the burning of fossil fuels causes greenhouse gas emission, which has many diver roles to the environment [1]. In addition, the world's oil supply is fixed since petroleum is naturally formed far too slowly in millions of years to be replaced at the rate at which it is being extracted [2]. Oil price also has increased over the years. As countries develop, industry and higher living standards drive up energy use, most often of oil. Thriving economies, such as India and China, are quickly becoming large oil consumers [3].

With the concern of depletion rate and price increase, there is rapidly growing interest in renewable energy source like biomass to be used as an alternative to fossil fuel. Biomass is biological material from living, or recently living organisms, most often referring to plants or plant-derived materials. Biomass can be converted to higher value products or energy. They include wide range of materials such as: virgin wood from forestry, energy crops specially grown for energy applications, agricultural residues from agriculture harvesting, food waste from food and drink preparation and processing, and post-consumer waste, or industrial waste and co-products from manufacturing and industrial processes [4].

The conversion technologies for utilization biomass can be separated into four basic categories: direct combustion processes, thermochemical processes, biochemical processes and

DOI: 10.2495/EQ-V3-N1-34-43

agrochemical processes [5]. Biomass conversion techniques including thermo-chemical and biochemical conversion are employed for power generation and production of liquid biofuels, chemicals and charcoal, which can be used as activated carbon and bio-energy.

Thermo-chemical conversion processes use heat or chemicals to convert biomass into another chemical form. Some of the basic techniques are: combustion, pyrolysis and gasification. These techniques are separated principally by the extent to which the chemical reactions involved are allowed to proceed. That is mainly controlled by the availability of oxygen and conversion temperature.

The coconut palm, *Cocos nucifera*, is a member of the family *Arecaceae* (palm family). It is the only accepted species in the genus *Cocos*. The conditions required for coconut trees to grow without any care are: mean daily temperature above 12°C–13°C (53.6–55.4°F) every day of the year [6]. Coconut palms are grown in more than 90 countries of the world, with a total production of 62 million tons per year [7]. Bangladesh has about 2777 hectares of coconut land and the average annual coconut production is about 450000 metric tons [8]. The husk and shells can be used for fuel and are a source of charcoal, which is an attractive source of syngas.

In the present study, we focused on the product yield and the gas formation behaviors during the pyrolysis of waste coconut shell and husk biomasses. The objective of this study is to gain a comprehensive understanding to the coconut shell and husk pyrolysis with focuses on the gas product releasing properties, thus facilitate to establish a universe model to simulate biomass pyrolysis based the coconut shell and husk compounds. It is favorable for the development of advanced biomass pyrolysis and gasification processes as bioenergy.

2 MATERIALS AND METHODS

2.1 Materials

In this experiment, coconut shell and husk were subjected to pyrolysis process. Coconuts were obtained from a local market of Jessore, located in Khulna division, Bangladesh. The shell and husk were separated from the coconuts and then the coconut shell and husk were sun dried for few days to remove the moisture content. After that dried coconut shell and husk samples were separately grounded and sieved to <250 μm by crusher (WB-1, 700 W, Osaka chemical, Co. Ltd., Japan) and sieve shaker (AS200 DIGIT 50 Hz, Retsch Co. Ltd., Japan), respectively, and then were dried at room temperature exposing 1 hour in the laboratory atmosphere for getting air-dried samples (JIS-M8811). Proximate analysis and ultimate analysis were performed on air-dried sample. Proximate analysis (Ash, volatile matter, moisture and fixed carbon) was conducted in accordance with the JIS-M8813. Ultimate analysis was done using a CHN corder (MT-5 Yanaco, Co. Ltd., Japan). The fundamental analyses of the coconut shell and husk are given in Table 1.

Table 1: Ultimate proximate and ultimate analyses of coconut shell and husk.

	Proximate analysis				**Ultimate analysis**			
[wt%]	M	A	V.M	F.C	H	C	N	O
Shell	7.82	0.23	79.91	12.04	4.5	39.2	0.2	55.9
Husk	9.96	2.23	72.60	15.21	5.4	48.95	0.4	43.1

2.2 Determination of chemical constituents

The major chemical element of coconut shell and husk were discerned. These include holocellulose, lignin and extractives. The chemical characteristics of coconut shell and coconut husk were determined in accordance with the standards outlined in wise methods. In the beginning step, extractives for example oils, waxes and resins were determined. In this process, a Soxhlet extraction flask of 250-ml capacity and Soxhlet extraction tube, inside diameter 30–40 mm, capacity to top of siphon about 100 ml, height of siphon tube about 55 mm were used. This type is specified because siphoning is more rapid than extractors with higher siphon tubes. The Soxhlet extraction flask was filled with 50 ml ethanol and 100 ml benzene mixture and heat 6 hours at 90°C using steam bath. Placed 2.0 to 10 grams of the air-dry sample into a tared extraction thimble. Placed the extraction thimble with specimen in clean and dry Soxhlet extraction apparatus. Connect the Soxhlet extraction apparatus in upright position to a clean and dry round bottom extraction flask. Connect the flask to the extraction apparatus and start water flow to the condenser section. Adjust the heaters to provide a boiling rate which will cycle the specimens for not less than 24 extractions over 6-hour period. Then removed the flask from the apparatus and partially evaporate the solvent in the extraction flask to a volume of 20–25 ml. Transfer the extract to the tared weighing dish by washing with small amounts of fresh solvent. Handle the weighing dish with forceps or tongs. Evaporate the solvent to near dryness while in the chemical fume hood. Dry the dish and contents in an oven for 1 hour at 105°C, cool in a desiccator, and weigh. Run a blank determination with the solvent used in the test. Evaporate 150 ml of the solvent to dryness, and weigh the residue. Correct the weight of the dried extract by the weight of residue found. Calculate the extractive content as follows equation (1).

$$\text{Extract, \%} = [(W_e - W_0)/W_b] \times 100 \tag{1}$$

where W_e = dry weight of extract, W_p = oven-dry weight of biomass, W_0 = dry weight of blank residue.

The determination of holocellulose was carried out in according to the wise method. In this process, take an Erlenmeyer flax and put 2.5 g extract removed sample. 150 ml deionized water, 1 g $NaClO_2$ and 0.2 ml acetic acid mixed with the sample. The flax was placed on a steam bath at 80°C with a 50 ml Erlenmeyer flax inverted over the top of the large flax and allowed to react. After 1 hour, 1 ml sodium chlorate and 0.2 ml acetic acid is added to the solution. The process was repeated more than three times. Then solution was filter and holocellulose where be submitted in the filter. Then the filter was dried over night after, then cooled for 1 hour and weighted. Calculate the holocellulose as flow equation (2).

$$\text{Holocellulose, \%} = [(H_e - H_0)/H_b] \times 100 \tag{2}$$

where H_e = dry weight of holocellulose, H_b = weight of biomass, H_0 = filter weight.

Lignin was analyzed using H_2SO_4 method. This method was performed with 1.00 g of air-dried extract free sample in 15 ml of 72% H_2SO_4 in a 50 ml flax. The reaction temperature was about 25°C and the biomass was thoroughly mixed into the 72% H_2SO_4 at the beginning. A glass rod was used and it was not removed from the flax until the end of the reaction stage. Kept this mixture for 4 hour in room temperature. After 4 hour mixture was placed in a 1000-ml tringle flax and 560 ml pure water was added. The solution was heated at 100°C for 4 h. After that the solution was cooled in room temperature and filtered using glass filter 1GP 16 and lignin was submitted in the filter. Then the filter was dried over night and after that cooled for 1 hour and weighted. Calculate the lignin as flow equation (3).

$$\text{Lignin\%} = [(L_e - L_0)/L_b] \times 100 \tag{3}$$

where L_e = dry weight of lignin, L_b = weight of biomass, L_0 = filter weigh.

2.3 Pyrolysis behavior of coconut shell and husk using TG-DTA

The pyrolysis of coconut shell and husk components were first carried using thermogravimetric differential thermal analyzer (TG-DTA). To mitigate the difference of heat and mass transfer, the sample weight was kept at 10 mg. The sample was heated up to 950°C at a constant heating rate of 10°C/min. Argon at a flow rate of 200 ml/min was used as the carrier gas to provide an inert atmosphere for pyrolysis and to remove the gaseous and condensable products, thus minimizing any secondary vapor-phase interactions. The conversion rate (x) was calculated from the weight loss measured using the TG-DTA and the initial weight. Weight loss was obtained from the conversion rate divided by the initial weight. The rate of the weight losses vs. time was expressed as dx/dt.

2.4 Experimental apparatus for coconut shell and husk pyrolysis

The design of the experimental setup used for coconut shell and husk pyrolysis and gasification is shown in Fig. 1. It was composed of gas feeding system, pyrolysis system, tar decomposition system, condensable products trapping system and gaseous products measurement system. Pyrolysis system and tar decomposition system consisted of the two stainless connecting reactors (inner diameter: 21.4 mm, length: 500 mm). Coconut shell and husk were placed on the mesh (40 μm) in each reactor. This apparatus can be independently heated in two different electric furnaces while the gas mixture coming in contact with the samples. Therefore, heterogeneous reaction between tar and ash can be carried out using the apparatus. The lines between the first and second reactors as well as between the second reactor and the condensable products trapping the system were heated at the temperature of around 300°C–400°C to avoid the condensation of tar. The condensable products were collected by cooling the test tube, and the collection capability was improved by using glass beads. The

a: Flow meter
b: Heating furnace
c: Temperature controller
d: Reactor
e: Biomass sample
f: Ribbon heater
g: Ice water cooler
h: Light tar and water trap
i: GC-TCD
j: GC-FID

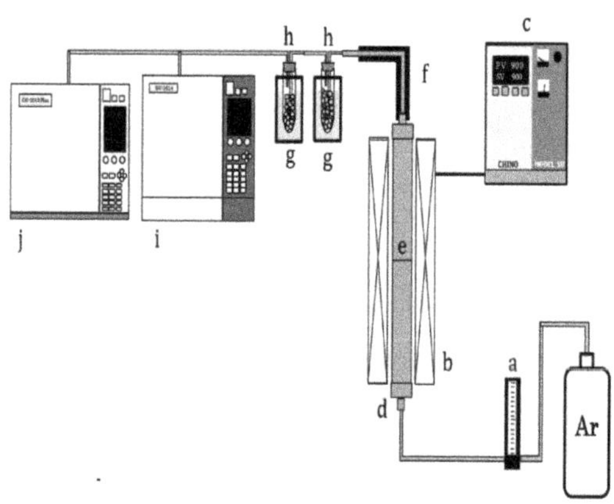

Figure 1: Experimental apparatus for pyrolysis products measurement.

cooling bath was kept below −3°C by mixing water and ice. The gaseous products were measured by a GC-TCD/FID (Model GC-2014; Shimadzu Co. Ltd., Japan). In this experiment, 10 kinds of gas were determined as gaseous products. H_2, CO, CH_4 and CO_2 were measured by a GC-TCD, while hydrocarbons (C_2H_6 and C_2H_4, C_3H_8, C_3H_6, iso-C_4H_{10}, n-C_4H_{10}) were measured by a GC-FID.

2.5 Measurement of pyrolysis product

Coconut shell and husk samples were heated under Ar atmosphere at a heating rate of 10°C/min from room temperature (about 25°C) to 900°C. The produced gases in the pyrolysis of coconut shell and husk samples were measured every 80°C from 100°C to 900°C. The produce of gaseous products (H_2, CO, CO_2, CH_4, C_2H_4, C_2H_6, C_3H_8, C_3H_6, n-C_4H_{10}, iso-C_4H_{10}) was calculated from the measurement results of GC-TCD (GC-2010, Shimadzu Co. Ltd., Japan) and GC-FID (Model GC-2014, Shimadzu Co. Ltd., Japan). The produce of tar was determined by the weight difference of the tar trap before and after the experiment. Tar contents are mainly present as 'light tar' and 'heavy tar' contents, considering that the light tar is collected in the tar trap and the heavy tar remained in the reaction tube. After completion of the experiment, open the electric furnace and cool the reaction tube for 1 hour using a fan. The produce of condensable product was determined by the weight difference of the tar trap before and after the experiment. After that, 20 ml of 2-propanol was injected into the test tube to dissolve the condensable product and the water content of the sample was quantified by the Karl Fischer moisture measurement method. The amount of char was weight that remaining in the inside after removal of the reaction tube.

3 RESULTS AND DISCUSSIONS

3.1 Holocellulose and lignin contents in coconut shell and husk

Cellulose and lignin are generally recognized as the main components in biomass. Holocellulose (cellulose and hemicellulose) and lignin were determined by chemical analysis for coconut shell and husk. Figure 2 showed the cellulose and lignin contents in coconut shell

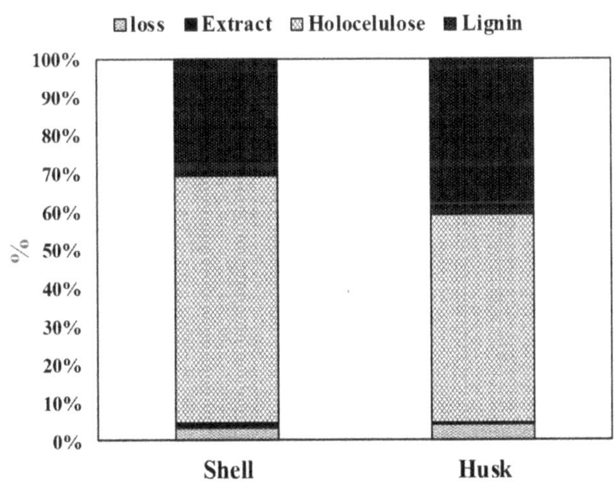

Figure 2: Cellulose and lignin contents in coconut shell and husk.

and husk. The term holocellulose is composed of cellulose and hemicellulose. The weight fraction except for the cellulose and lignin fraction corresponds to the fraction of acid-soluble hydrocarbons in biomass. As shown in Fig. 2, coconut shell and husk samples contain more cellulose than lignin. The cellulose and lignin contents vary from 54% to 65% and from 30% to 42%, respectively. The shell contains the highest amount of cellulose, up to about 65% and husk has a maximum of lignin, up to about 41%. Besides.

3.2 Pyrolysis behavior of coconut shell and husk using TG-DTA

Thermogravimetric analysis (TGA) records the loss of weight of a sample as the temperature is raised at a uniform rate. TGA of biomass samples has been extensively used as a means of determining the characteristics of the devolatilisation process. The TGA method provides information about net results of weight loss involving a range of complex chemical reactions in the thermal degradation of biomass. However, the data may be used to provide useful comparisons of the thermal degradation of different samples of biomass [9]. Thermal behavior of coconut shell and husk were studied by TG-DTA in the temperature range 20°C–950°C. Pyrolysis of biomass consists of three stages (Fig. 3). These stages can be outlined as: (i) dehydration process for temperatures below 122°C, (ii) pyrolytic cracking from 122°C to 400°C, stage consist of two exothermic simultaneous processes where the main components, hemicelluloses, cellulose and lignin are decomposed and a high amount of volatile matter formation occur and (iii) the last endothermic decomposition of the lignin at temperatures above 400°C. In first stage, as seen in Fig. 3 moisture removal stills up to 150°C. The ligno-cellulosic part of the coconut shell and husk is stable until 230°C with minor losses of moisture and low molecular weight volatile matter. In the second stage, main components of coconut shell and husk, hemicelluloses and cellulose, start degradation within the temperature range of 230°C–400°C. The weight loss was 60%, which was due to break down of hemicelluloses and cellulose and volatile matter removal. White *et al.* [10] studied that decomposition of cellulosic part of biomass occurs in two ways. In the first way, which occurs at low temperatures (up to 400°C in this study), break down of bounds in polymers and formation of CO, CO_2 and carbonaceous were occurred. In the second way, at high

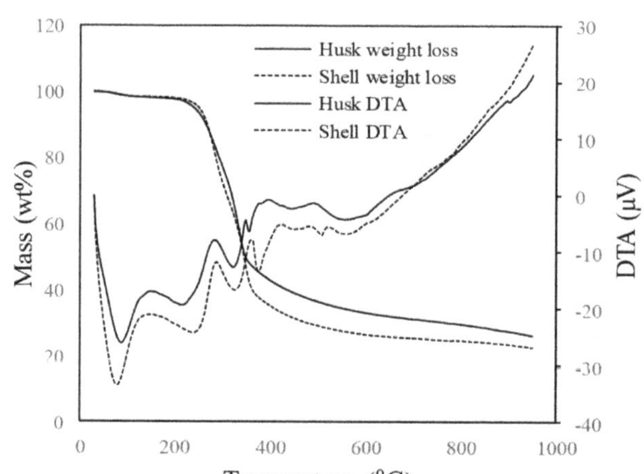

Figure 3: TG and DTA curves of coconut shell and husk.

temperatures (400°C–600°C in this study) integration of bonds, which leads to liquid formation, takes place. In the third stage, lignin was the most difficult one to decompose among the three main components, whose decomposition happened slowly under the whole temperature range from ambient temperature to 950°C, and the wide range of temperatures also caused the absence of its characteristic peak. Bilbao *et al.* [11] investigate that lignin volatilization occurs in a wide temperature range, which results a high char yield that maximizes at 460°C. Weight loss between 400 and 700°C can be attributed to decomposition of tars. After 700°C only a small change in weight loss has been observed.

3.3 Gases, tar and char yields obtained by pyrolysis

Figure 4 showed the yields of products (char, gas, water and tar) for coconut shell and coconut husk during the pyrolysis at 900°C. Coconut husk sample produced a much higher amount of char (34.6%) than the coconut shell (28.3%) during the pyrolysis. The result showed that more amounts of lignin produce more amounts of char, which was also observed at other biomass in the literature [12]. Among the biomass samples, coconut shell produced the largest amount of tar (25.5%) gas (26.3%). On the other hand, coconut husk produced less tar, which were 17.3%, respectively, and more amount of gas (26.9%). Nakorn Worasuwannarak *et al.* [13] also analyzed other biomass and they found that higher content of cellulose produced the largest amount of tar. Moreover, all the samples produced almost same amount of H_2O (i.e. 19.2% in coconut shell and 18.6% in coconut husk) during the pyrolysis. The much water content in the pyrolysis liquid product may be due to the release of volatile organic products during the preparation of condensed liquid tar sample. It is considerable that temperature conditions of the pyrolysis and gasification of tar are very important to reduce tar products for the pyrolysis and gasification processes [14–17] as bioenergy.

3.4 Product gases yields obtained by gas chromatography

The gas-evolving profiles from pyrolyzing the coconut shell and husk components in a packed bed, monitored by GC-TCD and GC-FID, are plotted in Figs 5 and 6. By comparison of Figs 5 and 6, almost similar points can be observed. H_2 was released out at a higher temperature (>450°C) and the releasing of H_2 increased greatly with temperature increasing. For coconut shell and coconut husk, the releasing of H_2 was significant and it got the maximum rate at 700°C after than the generation of H_2 was decrease in the

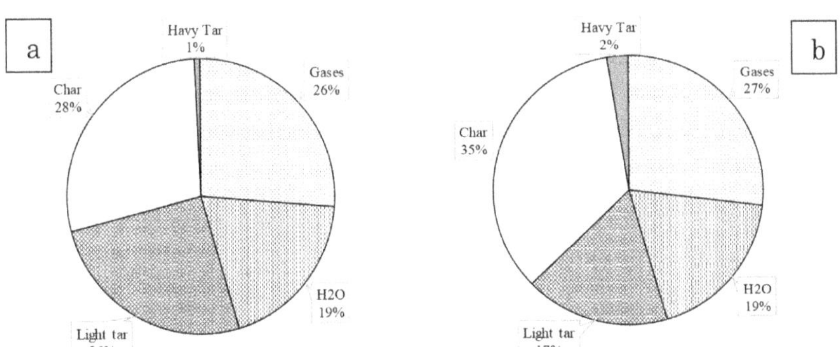

Figure 4: Pyrolysis products ratio of coconut (a) shell and (b) husk in Argon gas flow.

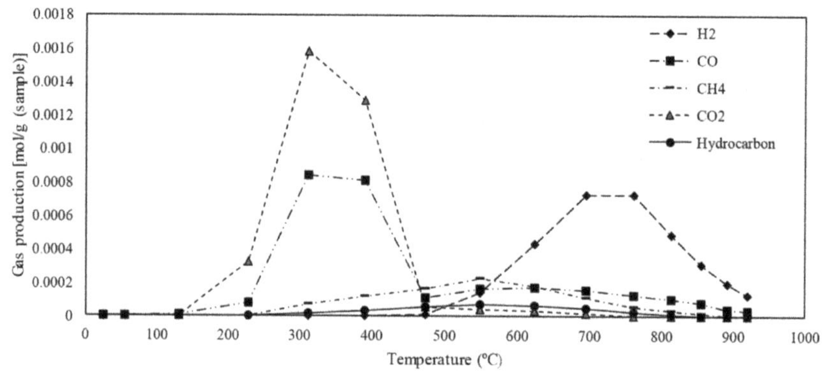

Figure 5: Releasing profile of gas products from coconut shell pyrolysis in Argon gas flow by heating at 10°C/min to 900°C.

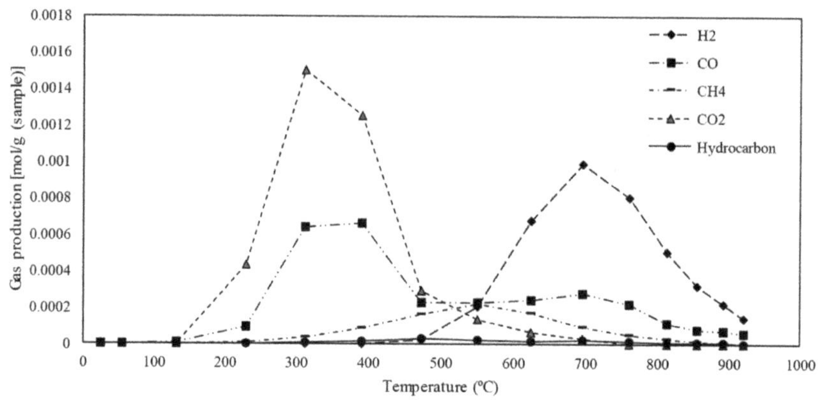

Figure 6: Releasing profile of gas products from coconut husk pyrolysis in Argon gas flow by heating at 10°C/min to 900°C.

temperature range. CO_2 was released out at 130°C–750°C and got the maximum releasing value at 300°C–400°C. The releasing of CO showed the almost similar pattern with that of CO_2. However, its releasing rate was lower than CO_2 and the maximum releasing rate of CO was found at 300°C–400°C. CH_4 was released out at the temperature range of 200°C–850°C and it got the maximum rate at 550°C. The releasing of hydrocarbon was generally very low.

The yield of gas species from pyrolyzing the coconut shell and husk components are listed in Table 2. It was calculated by integrating the curve of gas product evolving rate in the whole temperature range (25°C–900°C), thus it indicated the overall yield of each gas species from biomass pyrolysis. From the table, it can be observed that coconut shell owned the highest CH_4 yield; it might be attributed to the higher content of aromatic ring and O–CH_3 functional groups in the origin coconut shell sample. On the other hand, coconut husk owned the highest H_2. Production of H_2 from organics pyrolysis mainly came from the cracking and deformation of C=C and C–H, while CH_4 was mainly brought by the cracking of methoxyl. Moreover, coconut husk obtained the highest CO and CO_2 yield because of the higher carbonyl and carboxyl content.

Table 2: Yield of gas product from coconut shell and husk pyrolysis.

Sample	**Gas production yield (mg/g-sample)**				
	H_2	CO	CH_4	CO_2	Hydrocarbons
Coconut Shell	6.4	76.0	15.5	147.3	9.7
Coconut husk	7.8	81.3	13.7	162.5	3.6

4 CONCLUSION

In this study, pyrolysis experiments of the coconut shell and husk are carried out in TG-DTA and fixed bed reactor. The following are the findings from pyrolysis experiments of coconut shell in TG-DTA and fixed bed reactor:

1. The shell contains the highest amount of cellulose, up to about 65% and husk has a maximum of lignin, up to about 41% besides.
2. Main components of coconut shell and husk, hemicelluloses and cellulose, start degradation within the temperature range of 230°C–400°C. The weight loss was 60%, which was due to break down of hemicelluloses and cellulose and volatile matter removal.
3. Coconut husk sample produced a much higher amount of char (34.6%) than coconut shell (28.3%) during the pyrolysis. Among the biomass samples, coconut shell produced the largest amount of tar (25.5%). On the other hand, coconut husk produced less tar, which was 17.3%, respectively.
4. In the packed bed, CO_2 and CO mainly evolved out at 130°C–750°C, H_2 evolved out at a higher temperature (>450°C) and the yield of H_2 increased with pyrolysis temperature increasing. The releasing of H_2 was significant and it got the maximum rate at 700°C after than the generation of H_2 was decrease in the temperature range. It was suggested that different chemical structures of biomass components attribute to different gas product releasing properties. Coconut husk displayed a higher CO and CO_2 yield, mainly attributed to the thermal cracking of carbonyl and carboxyl content. With a higher presence of aromatic ring and methoxyl, the cracking and deformation of coconut shell released out much CH_4.

ACKNOWLEDGEMENT

Some works of this study was supported by the special funds for Basic Researches (B) (No. 15H05119, FY2015~FY2017) of Grant-in-Aid Scientific Research of the Japanese Ministry of Education, Culture, Sports, Science and Technology (MEXT), Japan.

REFERENCES

[1] Asadullah, M., Rahman, M.A., Ali, M.M., Rahman, M.S., Motin, M.A., Sultan, M.B. & Alam, M.R., Production of bio-oil from fixed bed pyrolysis of bagasse. *Fuel*, **86**, pp. 2514–2520, 2007.
https://doi.org/10.1016/j.fuel.2007.02.007

[2] Appenzeller, T., The end of cheap oil. *National Geographic*. June 2004.

[3] "China and India: A Rage for Oil", Business Week, 25 August 2005.

[4] Biomass Energy Center. Available at: http://www.Biomassenergycentre.org.uk

[5] Demirbas, A., Biomass resource facilities and biomass conversion processing for fuels and chemicals. *Energy conversion and Management*, **42**(11), pp. 1357–1378, 2001. https://doi.org/10.1016/s0196-8904(00)00137-0

[6] Chan, E. & Elevitch, C.R., *Cocos nucifera* (coconut) (version 2.1), 2006.

[7] Shashikumar, S. & Chandrashekar, H.M., An analysis of production and marketing of coconut in Tumkur District, India. *International Journal of Current Research and Academic Review*, **2**(10), pp. 167–175, 2014.

[8] Bangladesh Census of Agriculture, Department of Census and Statistics, 2011.

[9] Reed, A.R. & Williams, P.T., Thermal processing of biomass natural fibre wastes by pyrolysis. *International Journal of Energy Research,* **28**, pp. 131–145, 2004. https://doi.org/10.1002/er.956

[10] White, J.E., Catallo, W.J. & Legendre, B.L., Biomass pyrolysis kinetics: a comparative critical review with relevant agricultural residue case studies. *Journal Analytical and Applied Pyrolysis*, **91**(1), pp. 1–33, 2011. https://doi.org/10.1016/j.jaap.2011.01.004

[11] Bilbao, R., Mastral, J.F., Aldea, M.E. & Ceamanos, J., The influence of the percentage of oxygen in the atmosphere on the thermal decomposition of lignocellulosic materials. *Journal Analytical and Applied Pyrolysis*, **42**, pp. 189–202, 1997. https://doi.org/10.1016/s0165-2370(97)00050-8

[12] Burhenne, L., Messmer, J., Aicher, T. & Laborie, M-P., The effect of the biomass components lignin, cellulose and hemicellulose on TGA and fixed bed pyrolysis. *Journal of Analytical and Applied Pyrolysis*, **101**, pp. 177–184, 2013. https://doi.org/10.1016/j.jaap.2013.01.012

[13] Worasuwannarak, N., Sonobe, T. & Tanthapanichakoon, W., Pyrolysis behaviors of rice straw, rice husk, and corncob by TG-MS technique. *Journal Analytical and Applied Pyrolysis*, **78**, pp. 265–271, 2007. https://doi.org/10.1016/j.jaap.2006.08.002

[14] Wang, Q., Endo, T., Aparu, P. & Kurogawa, H., Study on biomass tar reduction by ash and fluidizing medium in a heterogeneous reaction. *International Journal of Sustainable Development and Planning*, **9**(5), pp. 669–679, 2014. https://doi.org/10.2495/sdp-v9-n5-669-679

[15] Wang, Q., Apaer, P., Kurokawa, H. & Sugiyama, K., Reactivity for pyrolysis and CO2 gasification of alkali metal loaded waste wood char. *International Journal of Sustainable Development and Planning*, **9**(5), pp. 680–691, 2014. https://doi.org/10.2495/sdp-v9-n5-680-691

[16] Wang, Q., Niida, H., Liu, C., Kurokawa, H., Sarkar, A., Sekiguchi, K. & Sugiyama, K., Oil aggregated behavior for coal recovery and combustion characteristics of their aggregates from different grade coals. *International Journal of Sustainable Development and Planning*, **9**(5), pp. 692–704, 2014. https://doi.org/10.2495/sdp-v9-n5-692-704

[17] Wang, W., Aparu, P., Kurogawa, H. & Sugiyama, K., Recovery briquetting technologies of waste biomass and pyrolyzed waste char produced from solid industrial and agricultural organic wastes. *International Journal of Sustainable Development and Planning*, **9**(5), pp. 705–716, 2014. https://doi.org/10.2495/sdp-v9-n5-705-716

IS DRY REFORMING THE SOLUTION TO REDUCE NATURAL GAS CARBON FOOTPRINT?

BRUNA REGO DE VASCONCELOS & JEAN-MICHEL LAVOIE
Industrial Research Chair on Cellulosic Ethanol and Biocommodities (CRIEC-B),
Department of Chemical and Biotechnological Engineering,
Université de Sherbrooke, Sherbrooke, Canada.

ABSTRACT

With the increasing world's energy demand along with the constantly expanding field of natural gas exploitation around the world, dry reforming of methane has gained increasing attention. Through this technology, natural gas can be converted into syngas, which is a well-known building block used for the production of alcohols and fuels. This technology has become an interesting approach for the valorization of a variety of CO_2 streams and for the reduction of the natural gas carbon footprint. In this work, attention will be given to the different reforming technologies used at industrial scale, followed by an investigation of the different approaches used for dry reforming of methane. Furthermore, focus will be given on how natural gas reforming could be used as a vehicle to store renewable energy while trying as well to reduce the carbon footprint of this technology. The technology presented in this work was previously developed by Hydro Québec and uses a cheap and available catalyst in addition to electricity to convert methane and carbon dioxide into syngas. Reactants conversions were up to 99% and the syngas produced had a H_2/CO ratio of 1 for over 200h.
Keywords: dry reforming, electricity, heterogeneous catalyst, natural gas, syngas.

1 INTRODUCTION

Following the COP21 conference in Paris in 2015, most of the world leading economies are seriously considering the possibility to reduce GHG emissions, which, in turn, requires lowering the utilization of fossil fuel. In Canada as an example, the federal government has a strong commitment in reducing GHG emissions throught an approach reported in the Pan-Canadian Framework on Clean Growth and Climate Change [1]. The main objective of this framework is to reduce by 2030 the country's GHG emission 30% below 2005 levels. This corresponds to about 525 Megaton eq. CO_2 per year for the whole country, which per capita corresponds to about 15 tonne per inhabitant. In order to reach this target, the approach will be to put a price on carbon and of course on CO_2 emissions either through a carbon tax or using a cap-and-trade system. As an example, the minimum value of carbon should be of 10 $(CAN) per tonne in 2018 moving up to 50$(CAN) per tonne in 2022. Canada has the advantage of having abundant natural resources including but not limited to biomass and hydroelectricity. This abundancy in natural resources has logically an impact on the price paid by the general public and the price of electricity in Quebec (one of Canada's province) is actually 0.0582 $/kWh (CAN) [2] for residential uses while it goes as low as 0.0327 $/kWh for industrial uses [3]. On the other side of the Atlantic Ocean, the European Union has committed towards reducing its GHG emissions 40% below 1990's level corresponding to 3,441 Megaton eq. CO_2 per year [4], which in turn represents 4.63 tonne per inhabitant per year. In 2015, the average price for electricity at household level was, in average, 0.22 €/kWh and about 0.125 €/kWh at industrial level [5].

Hence, despite the very variable commitments from the different government toward reducing GHG emissions, there is a general global tendency towards reduction of GHG emissions, which will require a significant reduction of fossil fuel utilization in order to materialize.

DOI: 10.2495/EQ-V3-N1-44-56

One possible approach to reach such a goal would be to replace large portions of fossil fuel consumption by renewable electricity or by renewable carbon-based substrates. As far as heating and cooling building is concerned, a careful utilization of renewable electricity is of course to be considered. However, some of the technologies allowing production of renewable electricity are constrained by peak periods during the day. Technologies such as solar or wind power are often more efficient during peak periods of the day making them a very variable and less reliable source to provide a steady energy source to large agglomerations. Hydroelectricity is of course a very stable source of renewable electricity although only available in large amounts in selected places around the world. Hence, in some cases, electricity is abundant and reliable, while for others it is a constant problem since it is necessary to cope for the periods during which electricity is less abundant (as an example during nighttime for solar energy). Another option would be to eventually cope for the lower electricity production period by using other technologies such as thermal plants operating on biomass CHP. However, biomass is not necessarily abundant everywhere and there is a strong necessity to secure a stable supply before relying on such option. Co-firing biomass with coal is another possibility although may cause some issues with regards to GHG emissions, especially if a reduction of carbon footprint is considered. Furthermore, adding more biomass to the electricity production may become especially challenging if transportation fuels also rely on this renewable carbon resource.

While the public and government's awareness and action towards global warming and GHG emissions increase on one side, the amount of fossil fuel being produced and consumed around the planet (and especially natural gas) is also significantly increasing. From 1990 to 2014, the world production of natural gas increased from 2457 to 4073 Billion cubic meter, which corresponds to an increase of about 166% in 30 years. Consumption of natural gas increased as well significantly over the last decades ranging from 52,944 Bcf in 1980 up to 122,804 Bcf in 2014 corresponding to an increase of more than a twofold in about 40 years. North America is the biggest consumer on the planet with 33,279 Bcf (about 25% of the consumption) followed by Asia and Oceania (24,737 Bcf). The latter has shown a very significant increase over the last 40 years ranging from 2,576 Bcf in the 80's up to a tenfold in the last couple of years. Among the different zones in the world, the only one that recorded a decrease in the last years was Europe while all the other regions of the world reported increase in consumption. As far as CO_2 emissions are concerned, natural gas consumption has led to the production of 6,983 MMTons CO_2 in 2014, which once again represents more than a twofold as compared to the production of CO_2 associated with the consumption of natural gas in 1980 [6].

Hence, there is on one side a significant interest and commitment from government and populations alike to favor sustainability and reduce GHG while the world's energy demand is increasing as well as fossil fuel exploitation. In many locations around the globe, natural gas has become one of the cheapest sources of carbon at this time (3.162 USD/GJ – May 2017) and with the actual reserves estimated to 6,879 Tcf in 2016, there is no doubt about the fact that this source for energy will be staying in the global portfolio for the years to come. Natural gas has been reported by many in open literature as a source of fossil carbon with a lower impact on carbon footprint [7, 8] but can this carbon balance be improved?

The world's energy is hence facing two problems and one of the best ways to address such a situation would be to combine both in order to create a new opportunity. On one hand, there is the abundancy of renewable energy that is either available at specific locations around the globe (such as in Quebec) or available in large amounts in specific time of the day such as wind and solar power. Such energy cannot, at this point, be efficiently stored in large scale in batteries although conversion to chemicals (such as hydrogen) may be considered. On the other hand, there is this wide interest towards natural gas that led to the constantly expanding field of natural gas exploitation around the world.

This work will hence focus on how natural gas reforming could be used as a vehicle in order to store renewable energy while trying to reduce the carbon footprint of such technology as well. Going through a process called Dry Reforming (DR), natural gas could be reacted with carbon dioxide originating from renewable carbon sources (such as first or second-generation ethanol processes) to produce syngas. The latter could downstream be used as a building block for fuels and chemicals. Such technology could as well be used as a storage system for renewable electricity since dry reforming is a highly endothermic reaction in addition to the fact that the syngas produced from this process has a low hydrogen to carbon monoxide ratio.

2 REFORMING TECHNOLOGIES

Environmental issues along with the increasing production and consumption of natural gas have raised questions on how to efficiently use this resource. Reforming of natural gas is a technology fully developed at industrial scale and used for syngas production. Among the reforming technologies, there are four main process to consider: steam reforming (SRM), partial oxidation (POX), autothermal reforming (ATR) and dry reforming (DRM). These technologies differ in the type of oxidizing agent used and in the H_2/CO ratio of the syngas produced.

Steam reforming of natural gas or steam reforming of methane (SRM) is the leading industrial process for natural gas reforming and for hydrogen and methanol production [9–11] and 80%–85% of the global hydrogen production is done through this process [9]. SRM is an endothermic reaction, which requires steam as oxidizing agent and produces a syngas with a H_2/CO ratio of 3 (eqn 1). This process is performed in multiple steps. In the first step, methane reacts with steam in two different reformers at temperatures between 800°C and 1000°C and at pressures around 15–30 atm. Despite being favored at low pressures, H_2 plants require the product at high pressures [12]. Catalysts employed in this step are mostly nickel-based catalysts. Then, the H_2/CO of the syngas produced in the first step is increased via water-gas shift reaction (WGS) (eqn 2) in two other reformers using an iron and copper-based catalysts. The products of the WGS reaction are then separated and purified. Under stoichiometric conditions (1 mol of water per mol of methane), carbon deposition is thermodynamically favored. So, steam is usually fed in excess with H_2O/CH_4 around 2.5–3. However, it has an impact on the overall cost of the process since steam generation requires significant amounts of energy [13].

$$CH_4 + H_2O \rightleftarrows CO + 3H_2 \quad \Delta H^{\circ}_{298K} = 206 \text{ kJ/mol} \tag{1}$$

$$CO + H_2O \rightleftarrows CO_2 + H_2 \quad \Delta H^{\circ}_{298K} = -41 \text{ kJ/mol} \tag{2}$$

Despite being a well-established technology, continuous improvement such as materials for reformer, catalyst performance and process concepts have resulted in reforming plants with lower investments and operational costs [12].

Partial oxidation of natural gas (POX) is an exothermic reaction that uses oxygen as oxidizing agent. In addition, such process usually requires a low energy investment when compared to SRM. In this process, methane reacts with oxygen to produce syngas with a H_2/CO ratio of 2 (eqn 3).

$$CH_4 + \frac{1}{2}O_2 \rightleftarrows CO + 2H_2 \quad \Delta H^\circ_{298K} = -38 \text{ kJ/mol} \tag{3}$$

POX can be performed catalytically with nickel-based catalysts or non-catalytically. In the latter case, it requires high temperatures (1300°C–1400°C) to achieve high conversions. Catalytical POX has the advantage of operating at lower temperatures and contact times, which implies small reactors with fast start-ups [14]. Nevertheless, the exothermicity of the combustion reaction limits the application of this process. Hot spots may be formed along the catalytic bed, which can provoke catalyst deactivation. Moreover, severe problems related to heat management and safety can also occur [12].

Authothermal reforming (ATR) is an approach resulting from the combination of SRM (eqn 1) and POX (eqn 3). First, natural gas is mixed with oxygen within a burner where POX takes place to provide heat for the reforming reaction. However, using pure oxygen would substantially increase the overall cost of the process hence, a mixture of oxygen, steam and/or CO_2 with methane is usually used [15]. Then, in the same reformer, steam and/or CO_2 react with the fuel remaining from the POX reaction to produce syngas. High temperatures between 1000°C and 1200°C are required to avoid the formation of byproducts. Therefore, part of the methane invested in the process is used for heating.

Dry reforming of natural gas or dry reforming of methane (DRM) has been considered as an alternative approach for syngas production from natural gas. This process (eqn 4) has gained increased attention since two main greenhouse gases, methane (CH_4) and carbon dioxide (CO_2), are converted into syngas at a H_2/CO ratio of 1 [12]. However, occurrence of simultaneous reverse water-gas shift reaction (RWGS) (eqn 5) may lead to a H_2/CO ratio lower than 1.

$$CH_4 + CO_2 \rightleftarrows 2CO + 2H_2 \quad \Delta H^\circ_{298K} = 247 \text{ kJ/mol} \tag{4}$$

$$CO_2 + H_2 \rightleftarrows CO + H_2O \quad \Delta H^\circ_{298K} = 41.2 \text{ kJ/mol} \tag{5}$$

Among the reforming technologies, DRM is the reaction requiring the largest amount of energy and oxidation of natural gas is the method usually employed at industrial scale to cover the energy requirements of the reaction. However, using renewable energy, such as electricity would decrease CO_2 emissions from the combustion of fossil fuel and would be a way of chemically store renewable energies, which is a challenge nowadays due to the intermittence of this type of energy.

Despite the environmental advantages and in contrast to the other reforming technologies, DRM is not fully applicable at industrial scale yet due to high energy requirements, fast deactivation of the commercial catalysts under the reaction conditions and requirement of a pure and steady source of CO_2 [15]. The CALCOR process (Caloric GmbH) and SPARG process (Haldor Topsøe) are two commercial technologies based on DRM reaction but are usually combined with SRM or POX in order to operate under milder conditions and to adjust H_2/CO ratios [12, 16].

3 REFORMING AND DRY REFORMING CATALYST

The most common way to lower the DRM energy demand to activate the reaction is the use of heterogeneous catalysts. However, the catalysts generally used in DRM suffers from fast deactivation mostly due to sintering in the high temperatures of reaction and carbon deposition via side reactions, such as methane cracking (eqn 6) and Boudouard reaction (eqn 7).

$$CH_4 \rightleftarrows C_{(s)} + 2H_2 \quad \Delta H^{\circ}_{298K} = 75\ kJ/mol \tag{6}$$

$$2CO \rightleftarrows C_{(s)} + CO_2 \quad \Delta H^{\circ}_{298K} = -171\ kJ/mol \tag{7}$$

Transition metals-based catalysts (Ni, Fe, Co …) and noble metals-based catalysts have been investigated for this reaction. Noble metals-based catalysts such as Pt, Ru, Rh and Pd have shown to be highly active in DRM reaction with good resistance against carbon formation. Rostrup-Nielsen *et al.* [17] compared the activity of different noble metals (Rh, Ru, Pt and Pd) and Ni supported on MgO. They reported the following order of performance: Ru > Rh > Ni > Ir > Pt > Pd. They showed that the rate of carbon formation was much lower for the noble metals-based catalyst than for the Ni catalyst, especially at 500°C, temperature at which carbon forming reactions are favored. Similarly, Coronel *et al.* [18] tested 0.6 wt% Rh/La_2O_3–SiO_2 catalyst in the DRM reaction for over 180 h of time on stream. The catalyst was highly active and stable and no carbon formation was detected. Ballarini *et al.* [14] investigated the performance of a platinum catalyst with metal loadings varying between 0.02, 0.1 and 0.5 wt.%. They observed that methane conversion increased with platinum content. After 135 min of reaction, significant deactivation was observed for the catalysts containing 0.02 and 0.1 wt%Pt. However, analysis of the spent catalysts showed negligible carbon contents, suggesting that the deactivation was provoked by sintering of the platinum particles.

Despite their good performance, the costs of noble metals catalysts are much higher than the transition metals catalysts, which hinders their utilization at large scale. In addition, transition metals-based catalysts, especially nickel, have a history in the field of reforming and especially for SRM processes. Li *et al.* [19] investigated the effect of Ni contents from 2 to 20 wt.%. They showed that the performance of the catalyst gradually increased with Ni loadings from 2 to 10 wt.%. However, further increase led to loss in catalytic stability mostly due to increase in the size of nickel particles and to carbon deposition. Similarly, Kroll *et al.* [20] investigated the performance of Ni/SiO_2 catalyst on DRM at 700°C. The catalyst showed an initial methane conversion around 80%, that decreased to about 70% in the first hours of reaction and remained stable for 80 h of time on stream. The initial deactivation was related to nickel particles sintering and coke deposition.

So, the longevity of the catalyst has become a crucial parameter in the approach of using heterogeneous catalysts to lower the DRM energy demand to activate the reaction. In order to improve coke resistance of transition metals-based catalysts, dopants to increase basicity and consequently carbon removal and improve metal dispersion over the support have been investigated. Therdthianwong *et al.* [21] tested the effect of ZrO_2 on the performance of Ni/Al_2O_3 catalyst. They reported that ZrO_2 addition enhanced the dissociation of CO_2 forming oxygen intermediates, which improved gasification of coke deposits. Similarly, Alipour *et al.* [22] investigated the influence of alkaline and alkaline-earth promoters such as BaO, CaO and MgO on the performance of Ni/Al_2O_3 catalyst during DRM at 700°C. They reported that the addition of the promoters enhanced the basic properties of the catalysts and consequently favored CO_2 adsorption, which resulted in higher coke resistance.

4 USE OF RENEWABLE ELECTRICITY – THE HQ–UDES PROCESS AND SCALE-UP

Many approaches have been reported so far in literature when it comes to dry reforming of methane or short chain alkanes. Many of the researches published in open literature reports on the utilization of classical nickel-based catalyst in a fixed bed reactor. However, other reports aim at the utilization of different system and in some cases, involving electricity.

Tu *et al.* reported on the dry reforming of methane using a classical nickel on gamma-alumina catalyst but with a plasma [23]. For the experiments, the authors developed a coaxial dielectric barrier discharge (DBD) reactor that they operated at atmospheric pressure. The results showed that a combination of the higher power level used in this experiment combined with the lower debit allowed an optimal conversion of methane and carbon dioxide into syngas. The gas flow used in this work however was of 50 mL/min (25 mL of both reactant) for an optimal power requirement of 50 W. Hence the power requirement is about 3 times the total power value of the methane being processed through the reactor which, combined to the complexity of the custom-made plasma reactor may inhibit the possibility to scale such technology at larger scale.

Sekine *et al.* also investigated the effect of electricity on the dry reforming reaction using a non-equilibrium electric discharge at ambient temperature [24]. In this specific case, the interest was for biogas, which in most cases has a composition comparable to what is required for state of the art dry reforming reactions. Another important factor of this work is that the reaction was performed at ambient temperature, which is quite uncommon for dry reforming reactions that are known to operate at high temperatures. The system was found very effective for H_2S conversion at low power input. As far as methane conversion was concerned, the work reported a total of 1.2 W power input for a total conversion of 60% of methane. If compared to the previous work, the total flow rate for the biogas was 10 cc/min; hence, in this case the power required for the conversion of biogas was 37% the calorific value of the methane injected in the reactor leading to a conversion of a little more than half the methane being injected in the reactor. Despite the fact that the system, as the one reported by Tu *et al.*, may be hard to scale-up, there is nevertheless an undeniable potential for conversion of H_2S that reached in some cases close to 90%. For this specific work, no catalyst was reported and the composition of the electrodes was not specified to the best of our knowledge.

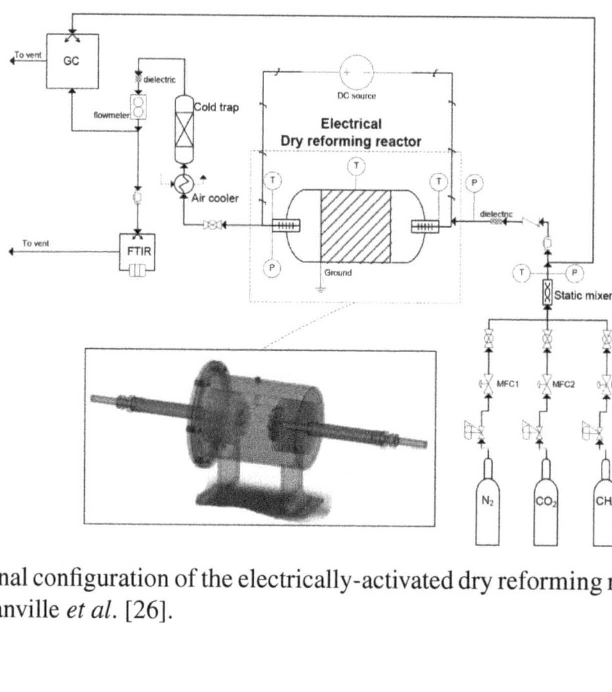

Figure 1: Original configuration of the electrically-activated dry reforming reactor as reported by Banville *et al.* [26].

Another conversion approach that was reported in literature for dry reforming of methane was reported by Labrecque and Lavoie in 2011. In this work, the authors claimed that the utilization of an electrified fixed bed of steel wool allowed high conversions both of methane and of carbon dioxide [25]. This approach could possibly lead to new and disruptive innovations in the field since it used a catalyst that was not extensively reported in that field before and a reactive system that was overall complex although simpler and cheaper to scale than the plasma-based systems. The authors reported on a 343 W consumption of electrical current for an overall conversion close to 70% of the methane and CO_2 entering the reactor. However, the feed of methane was reported to be 1.70×10^{-4} mol/s of methane, which makes the power required for the conversion significantly higher than the amount of energy originally fed in the reactor (via the calorific value of methane). Although at this point energetically unfavorable, the concept behind de reactor was slightly simpler with a common catalyst (steel wool bought at the hardware store and sandwiched between two electrodes (Fig. 1).

Later, the same authors reported on another system conceptually quite comparable to the one reported in Fig. 1 except for the outside design of the reactor that was this time made of

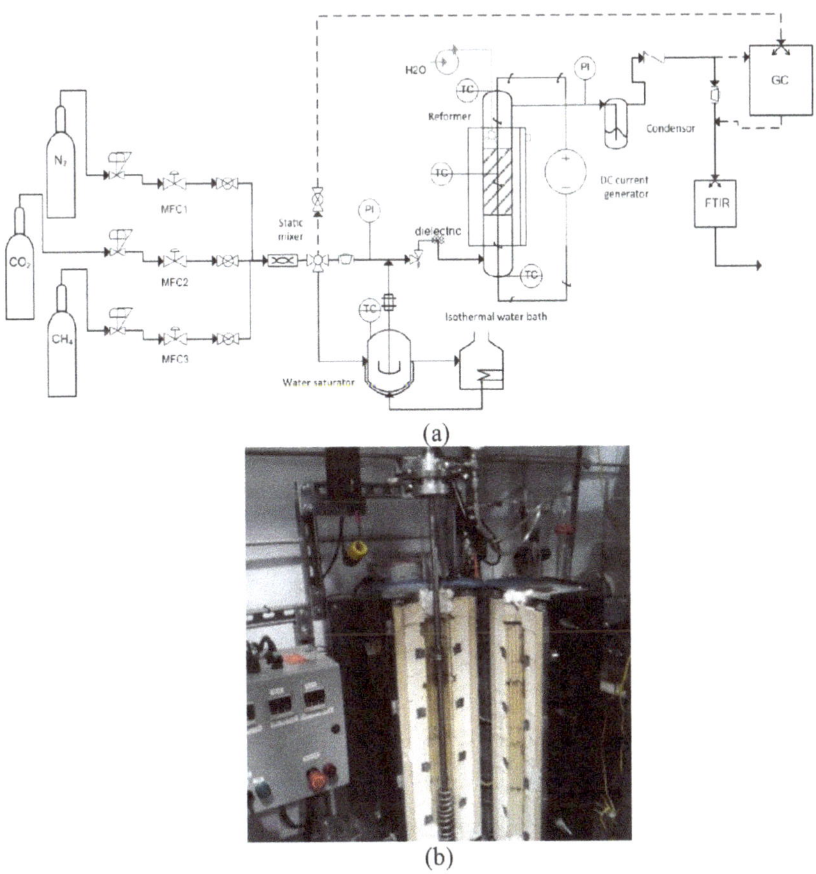

Figure 2: Configuration of the second generation of the electricity-activate dry reforming reactor where (**A**) is the P&ID of the overall system and (**B**) shows a picture of the reactor.

quartz instead of steel. In addition, the reactor in this case was also externally heated using a tubular furnace contrarily to the system reported in Fig. 1. Figure 2A shows the general P&ID of the modified reactor [27] while Fig. 2B shows a picture of the system.

From the second generation of catalyst and following an optimization of the different parameters, the team was able to reduce the amount on energy invested in the system while increasing the conversion both of methane and carbon dioxide. Hence, the team was able to reach total conversion of both reactant with a power investment of 65W using a GHSV of 3200 h^{-1} corresponding to a total inlet flow of about 1 L/min at 950°C. In terms of energy requirement, the amount of electricity invested in this case corresponded then to about 20% of the value of methane injected in the reactor. While reducing the amount of electricity invested, the team pursued their work on this technology and are now completing the

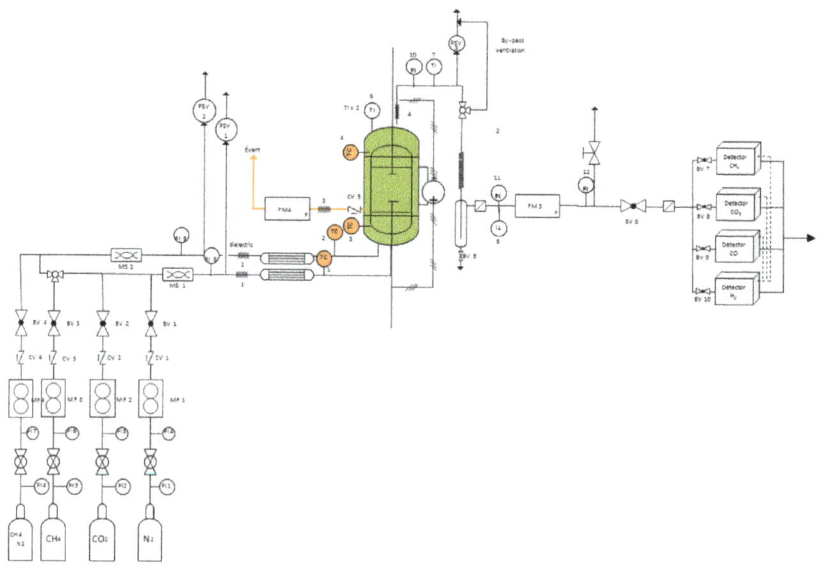

(a)

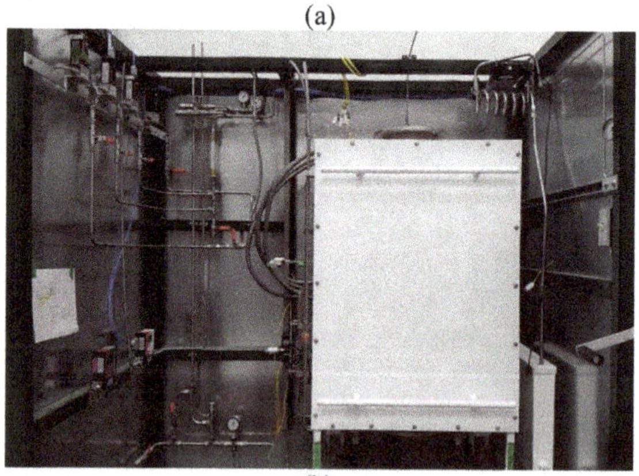

(b)

Figure 3: Configuration of the pilot scale reactor with a capacity of 40 L/min where (**A**) is the P&ID of the overall system and (**B**) shows a picture of the reactor.

assembly of a pilot scale reactor with a capacity of 40 L/min. The technology now relies on one patent and one patent pending process and should be compatible for natural gas, biogas and tail gas. Figure 3 depicts the P&ID of the scaled-up system and a picture of the most recent advances.

5 MASS & ENERGY BALANCE – ECONOMICAL POTENTIAL

As mentioned earlier, dry reforming is a process by which carbon dioxide is reacted with methane in order to produce two moles of carbon monoxide as well as two moles of molecular hydrogen (eqn 4). Using dry reforming instead of classical reforming will involve sequestration of 1 mole of CO_2 per mole of methane reacted. Since carbon dioxide has a higher molar mass than methane and it is required to react 1 mole of methane per mole of carbon dioxide, a tonne of carbon dioxide will permit conversion of about 364 kg of methane. If this reality is considered from the methane perspective, it means that, in order for the dry reforming process to be efficient at a 1/1 ratio between the two gases it will require 2.7 metric tonne of CO_2 to react with one tonne of methane. Hence dry reforming of one tonne of methane could lead to the sequestration of almost 3 tonnes of carbon dioxide.

In Canada, carbon dioxide is expected to worth in the range of 50$ per tonne in 2022 [1]; hence, utilization of CO_2 in a dry reforming process could provide additional benefits of 135$(CAN) per tonne of methane converted. In provinces such as Quebec (Canada), the price of electricity for industrial application is of 0.0327$(CAN) per kWh (hence for 3600 kJ). As mentioned earlier, dry reforming is an endothermic process that requires in the range of 247.3 kJ/mol to be performed. Hence, each tonne of methane converted should require 4214 kWh to be converted to syngas, which would overall represent at the above reported price a cost of about 138$(CAN) per tonne of methane converted using dry reforming, which is about the same amount of money that would be generated from the carbon credits on CO_2.

If such situation was to be compared with steam reforming of methane, and assuming a scenario where the price of methane would remain the same (2.85 USD/GJ – June 2017), the first target would be to assess on the price required for the conversion of one tonne of methane considering the endothermicity of this reaction as well. The steam methane reforming has an endothermic value of 206.8 kJ/mol and the calorific value of methane is 55530 kJ/kg. Hence, one tonne of methane (55.7×10^7 kJ) would require 1.29×10^7 kJ for the endothermic part of the steam reforming reaction, which correspond roughly to 23.2% of the energy value of methane dedicated to cope for the thermodynamics of the reaction. This would overall represent an investment of 12.9 GJ per tonne of methane converted, which, according to the prices mentioned above, would represent an investment of 49.53$(CAN) to convert one tonne of methane. Hence, in 2022 if both electricity and methane remains at the same price as they are right now and that the price of CO_2 have reached 50$(CAN) per tonne, conversion of natural gas with a combination of renewable CO_2 and electricity would be more economical than classical steam reforming.

In order to break even with the actual price of natural gas and using renewable electricity from Quebec at industrial rate, a break-even scenario between the price invested for the thermodynamics of both process would be when the price of CO_2 would be in the range of 20–30$(CAN) per tonne, which is envisioned to be in 2020–2021 [1]. In such scenario, the endothermic penalty of dry reforming would cost about the same as the one for steam methane reforming. All these calculations remain speculations but overall, it is clear that there is progressively room for processes such as dry reforming and that on the strict economical point of view, there might be in a near future and in specific location around the globe (such as in Quebec) the possibility to get such technologies at industrial level. Of course, other considerations will have to be considered, such as the reduction of the carbon footprint of dry reforming technologies as well as being able to produce a new 50/50 renewable/fossil variety of fuels and chemicals (or 75/25 on mass basis).

6 CONCLUSIONS

Dry reforming technologies could be a very disruptive technology that could allow solving two problems by combining them into a sole opportunity. With the recent withdrawal of the US from COP21 (June 2017), and since the US is one of the biggest users of natural gas around the planet, there is no doubt that this form of fossil fuel will be sticking around for the upcoming decades. Reforming can be performed with different kind of oxidants ranging from the highly exothermic reactions obtained with air or pure oxygen to the endothermic reactions obtained when using steam and carbon dioxide. While steam methane reforming offers the advantage of producing a hydrogen-rich syngas, dry reforming (that uses CO_2 as oxidant) has the advantage of using another GHG, and depending from the source of this carbon dioxide, such an approach could be very beneficial in a world where carbon footprint is becoming a global concern. Using carbon dioxide, as an example, from first or second biofuel process could lead to a net reduction of natural gas carbon footprint and overall decrease the impact related to this fossil fuel and its ever-expanding exploitation and market. Since dry reforming involves an equimolar reaction between carbon dioxide and methane and assuming a renewable source of CO_2 (such as CO_2 from fermentation) the overall carbon footprint of methane on a strict molecular footprint would decrease by 50%.

Still in accordance with the objectives of COP21, one of the main challenges on a global scale is to reduce the production of electricity being produced from fossil fuel. Hence, production of renewable electricity can be done using different process such as (but not limited to) wind power, solar power, biomass cogeneration and hydroelectricity. In many cases however there is either a punctual excess or an overall excess of such energy and always remains the challenge of storing this energy in light of future utilizations. From this reality has developed the concept of e-fuels where renewable electricity is used to transform CO_2 into fuels or chemicals (such as methane of methanol). In reactions such as reforming (steam or dry) both reactions are very energy intensive in light of very endothermic processes. In order to cope for the system's demand in energy, renewable electricity could be used to heat reactions hence coping for this thermodynamic penalty. In common reforming systems such as steam reforming technologies, the energy requirement of the reaction is covered by the utilization of part of the natural gas (or methane) that is fully oxidized to produce the energy required by the process. Hence, by using electricity it allows reducing combustion of fossil fuel and overall the carbon footprint of this technology.

Combining both renewable CO_2 and renewable electricity in dry reforming process would overall significantly reduce the carbon footprint of natural gas exploitation and utilization in the field of fuels and chemicals synthesis. In this work was reported the first process relying on the combined utilization of renewable carbon dioxide and electricity. The UdeS process allows conversion of methane and carbon dioxide up to 99%+ into a syngas with a H_2:CO ratio of 1. This technology, originally developed by Hydro Quebec, relies on a cheap and available catalyst (Bulldog steel wool) and has been scaled from laboratory to pilot scale. Such technologies, allying the concept of e-fuels to fossil fuel, could represent a game changer in the upcoming years since they allow renewable energy storage combined to positively impacting the carbon footprint of natural gas. However, dry reforming also leads to a low hydrogen content as compared to most of the reforming technologies on the market. It is nevertheless possible to increase the hydrogen content in the syngas by adding renewable hydrogen produced from the electrolysis of water into the process, which could allow displacing again fossil fuels. It now seems clear that many solutions are available and the upcoming challenges for engineers and scientist will be to take such concept and make them reach a full industrial scale. Costs will necessarily be an issue, although carbon credits combined with a strong determination from governments to push forward GHG reduction technologies to make of this era one of the most suitable for changes.

ACKNOWLEDGEMENTS

The authors would like to acknowledge contribution from the BioFuelNet network (Gasification project work package) as well as MITACS for Dr Bruna Rego de Vasconcelos postdoctoral fellow grand. The authors would also like to acknowledge the original sponsors of this work (Enerkem, Hydro Quebec and CRB Innovations) as well as NSERC. Finally, the authors would like to acknowledge the sponsors of the Industrial Research Chair on Cellulosic Ethanol and Biocommodities for their contribution to this work.

REFERENCES

[1] Government of Canada, PAN-CANADIAN FRAMEWORK on Clean Growth and Climate Change, 2017.

[2] Hydro Québec, Tarif D - Tarif domestique pour la clientèle résidentielle et agricole, (n.d.). Available at: http://www.hydroquebec.com/residentiel/espace-clients/compte-et-facture/comprendre-facture/tarifs-residentiels-electricite/tarif-d.html.

[3] Hydro Québec, Tarif L - Tarif industriel pour la clientèle de grande puissance, (n.d.). Available at: http://www.hydroquebec.com/grandesentreprises/tarification/tarifs-grande-puissance/tarif-l/ (accessed June 6, 2017).

[4] European Commission, Eurostat - Statistics Explained. Total greenhouse gas emissions by countries, 1990-2014. (n.d.). Available at: http://ec.europa.eu/eurostat/statistics-explained/index.php/File:Total_greenhouse_gas_emissions_by_countries_(including_international_aviation_and_indirect_CO2,_excluding_LULUCF),_1990_-_2014_(million_tonnes_of_CO2_equivalents)_updated.png (accessed June 6, 2017).

[5] European Commission, Eurostat - Statistics Explained. Electricity and gas prices, second half of year, 2013–2015 (n.d.). Available at: http://ec.europa.eu/eurostat/statistics-explained/index.php/File:Electricity_and_gas_prices,_second_half_of_year,_2013–15_(EUR_per_kWh)_YB16.png (accessed June 6, 2017).

[6] U.S. Department of Energy, International Energy Statistics, (n.d.). Available at: https://www.eia.gov/beta/international/data/browser/index.cfm#/?pa=000000000000000000004&tl_id=3002-A&vs=INTL.3-6-AFRC-TCF.A&cy=2014&vo=0&v=H&start=1980 (accessed June 6, 2017).

[7] U.S. Department of Energy, International Energy Outlook 2016 - Chapter 3. Natural gas, 2016. Available at: https://www.eia.gov/outlooks/ieo/nat_gas.cfm (accessed June 6, 2017).

[8] ICF Consulting Canada, Life Cycle Greenhouse Gas Emissions of Natural Gas - A Literature Review Of Key Studies Comparing Emissions From Natural Gas and Coal, pp. 1–8, 2012. Available at: http://www.capp.ca/getdoc.aspx?DocId=215278.

[9] Alhamdani, Y.A., Hassim, M.H., Ng, R.T.L. & Hurme, M., The estimation of fugitive gas emissions from hydrogen production by natural gas steam reforming. *International Journal of Hydrogen Energy,* **42**(14), pp. 9342–9351, 2017. https://doi.org/10.1016/j.ijhydene.2016.07.274

[10] Anzelmo, B., Wilcox, J. & Liguori, S., Natural gas steam reforming reaction at low temperature and pressure conditions for hydrogen production via Pd/PSS membrane reactor. *Journal of Membrane Science*, **522**, pp. 343–350, 2017. https://doi.org/10.1016/j.memsci.2016.09.029

[11] Johnson Matthey, Reforming (ATR, GHR, SMR), (n.d.). Available at: http://www.jmprotech.com/core-technologies-reforming-ATR-GHR-SMR.

[12] Ferreira-Aparicio, P., Benito, M.J. & Sanz, J.L., New trends in reforming technologies: from hydrogen industrial plants to multifuel microreformers. *Catalysis Reviews*, **47**(4), pp. 491–588, 2005. https://doi.org/10.1080/01614940500364958

[13] Darujati, A.R.S. & Thomson, W.J., Kinetic study of a ceria-promoted catalyst in dry-methane reforming. *Chemical Engineering Science*, **61**(13), pp. 4309–4315, 2006. https://doi.org/10.1016/j.ces.2006.02.007.

[14] Ballarini, A., Basile, F., Benito, P., Bersani, I., Fornasari, G., De Miguel, S., Maina, S.C.P., Vilella, J., Vaccari, A. & Scelza, O.A., Platinum supported on alkaline and alkaline earth metal-doped alumina as catalysts for dry reforming and partial oxidation of methane. *Applied Catalysis A: General,* **433–434**, pp. 1–11, 2012. https://doi.org/10.1016/j.apcata.2012.04.037

[15] Lavoie, J.M., Review on dry reforming of methane, a potentially more environmentally-friendly approach to the increasing natural gas exploitation. *Frontiers in Chemistry*, **2**, pp. 1–17, 2014. https://doi.org/10.3389/fchem.2014.00081

[16] Jafarbegloo, M., Tarlani, A., Mesbah, A.W. & Sahebdelfar, S., Thermodynamic analysis of carbon dioxide reforming of methane and its practical relevance. *International Journal of Hydrogen Energy*, **40**(6), pp. 2445–2451, 2015. https://doi.org/10.1016/j.ijhydene.2014.12.103

[17] Rostrup-Nielsen, J.R. & Hansen, J.H.B., CO2 reforming of methane over transition metals. *Jouranl Catalysis*, **144**(1), pp. 38–49, 1993. https://doi.org/10.1006/jcat.1993.1312

[18] Coronel, L., Múnera, J.F., Lombardo, E.A. & Cornaglia, L.M., Pd based membrane reactor for ultra pure hydrogen production through the dry reforming of methane. Experimental and modeling studies. *Applied Catalysis A: General*, **400**(1–2), pp. 185–194, 2011. https://doi.org/10.1016/j.apcata.2011.04.030

[19] Li, L., Zhang, L., Zhang, Y. & Li, J., Effect of Ni loadings on the catalytic properties of Ni/MgO(111) catalyst for the reforming of methane with carbon dioxide. *Journal Fuel Chemistry and Technology,* **43**(3), pp. 315–322, 2015. https://doi.org/10.1016/S1872-5813(15)30007-4

[20] Kroll, V.C.H., Swaan, H.M., Lacombe, S. & Mirodatos, C., Methane reforming reaction with carbon dioxide over Ni/SiO2Catalyst. *Journal of Catalysis*, **164**(2), pp. 387–398, 1997.

[21] Therdthianwong, S., Siangchin, C. & Therdthianwong, A., Improvement of coke resistance of Ni/Al2O3 catalyst in CH4/CO2 reforming by ZrO2 addition. *Fuel Processing Technology*, **89**(2), pp. 160–168, 2008.
https://doi.org/10.1016/j.fuproc.2007.09.003

[22] Alipour, Z., Rezaei, M. & Meshkani, F., Effect of alkaline earth promoters (MgO, CaO, and BaO) on the activity and coke formation of Ni catalysts supported on nanocrystalline Al2O3 in dry reforming of methane. *Journal of Industrial Engineering Chemistry*, **20**(5), pp. 2858–2863, 2014.
https://doi.org/10.1016/j.jiec.2013.11.018

[23] Tu, X. & Whitehead, J.C., Plasma-catalytic dry reforming of methane in an atmospheric dielectric barrier discharge: understanding the synergistic effect at low temperature. *Applied Catalysis B: Environmental*, **125**, pp. 439–448, 2012.
https://doi.org/10.1016/j.apcatb.2012.06.006

[24] Sekine, Y., Yamadera, Y., Matsukata, M. & Kikuchi, E., Simultaneous dry reforming and desulfurization of biomethane with non-equilibrium electric discharge at ambient temperature. *Chemical Engineering Science*, **65**(1), pp. 487–491, 2010.
https://doi.org/10.1016/j.ces.2009.06.011

[25] Labrecque, R. & Lavoie, J., Dry reforming of methane with CO2 on an electron-activated iron catalytic bed. *Bioresource Technology*, **102**, pp. 11244–11248, 2011.
https://doi.org/10.1016/j.biortech.2011.09.088

[26] Banville, M., Lee, R.A., Labrecque, R. & Lavoie, J.M., Interaction of CO2/CH4 with steel wool in an electrocatalytic dry reforming reactor. *Energy and Sustainability IV*, **176**, pp. 17–28, 2013.
https://doi.org/10.2495/ESUS130021

[27] Banville, M., Labrecque, R. & Lavoie, M.J., Dry reforming of methane under an electro-catalytic bed: effect of electrical current and catalyst composition. *Energy and Sustainability V*, **186**, pp. 603–611, 2014.
https://doi.org/10.2495/ESUS140531

DGNB BUILDING CERTIFICATION COMPANION: SUSTAINABILITY TOOL FOR ASSESSMENT, PLANNING, LEARNING, AND ENGAGING (STAPLE)

RENATE SKOVGAARD MØLLER, MICHAEL K. RHODES & TINE STEEN LARSEN
Aalborg University, Department of Civil Engineering, Denmark.

ABSTRACT

In the construction industry, the popularity of sustainability and its benefits have been on the rise in recent years. Alas, with various building sustainability assessment schemes on the market, there is still no single general method for a comprehensive and inclusive design and building process for sustainable buildings. The literature describes several barriers of entry preventing actors in the industry from seeking sustainability certifications and prioritizing design methods, supporting sustainability in greater numbers. In the newly developed tool, "DGNB building certification companion: Sustainable Tool for Assessment, Planning, Learning, and Engaging (STAPLE)", a new Excel-based, interactive, and iterative education focused platform is introduced, intended to engage dialog among stakeholders, building owners, and decision makers, and the assigned group team leaders, based on the five DGNB topics. In order to establish common levels of knowledge, terminology, and understanding for proper interdisciplinary discussions, which would result in suitable and timely decisions, personal and professional development is enabled by imbedded educational documents in multiple formats throughout the tool as plain-language, easily digestible summaries of various topics regarding sustainability and the DGNB certification scheme. The identified barriers are described in the tool followed by a solution to overcome them. The tool, tested at multiple stages of development and moulded by many individuals both within and outside of the sustainable building industry, has shown to achieve the primary goals of assessment of individual's current knowledge, educating through multiple stages and formats, and the inspiring of conversation among team members through a graphical display of opinions. Based on user feedback, the conclusion was that this is a desired product on the market. This new approach is expected to dramatically reduce misunderstandings, conflicts, and mistakes during a sustainable design process, helping the design team plan a project to possibly obtain the highest DGNB score if desired and properly documented.

Keywords: building sustainability assessment schemes, DGNB, integrated design process, sustainability, sustainability certifications, sustainability certification companion tool.

1 INTRODUCTION

Rooted in the belief that: '*To pursue sustainability is to create and maintain the conditions under which humans and nature can exist in productive harmony to support present and future generations.*' [1], sustainability remains a priority interest of many organizations and groups around the world. As a significant contributor to pollution and resource consumption [2], the global building and construction industry faces several challenges to maintain the same level of quality and keep up with the modern day demands, whilst living up to the worldwide sustainability aspirations.

There are several initiatives and strategies to improve these issues both globally and locally, such as the development of new building standards. In Denmark, the Building Regulations are relatively stringent, encouraging healthy indoor environment, lowered energy consumption, and safe building practices [3]. Sustainable building certifications and rating systems is another method to achieve lowered negative environmental, economic, and social impact by promoting and encouraging sustainable design. While there are multiple sustainable building certifications around the world, DGNB, developed by *Deutsche Gesellschaft für*

DOI: 10.2495/EQ-V3-N1-57-68

Nachhaltiges Bauen, or in English, *German Sustainable Building Council*, is the process often used in Europe, and adapted and modified in Denmark to match the current national building regulations [4]. However, it is worth noting that DGNB remains a voluntary option and is not part of the current building regulations in Denmark.

DGNB is a sustainable building assessment system, based on a weighted list of indicators, which, if completed through building design, award points in various categories, if properly documented and awarded by a third-party auditor. Sustainable design principles are promoted such as *Responsible procurement* and *Waste management*. Uniquely, economy is viewed as a resource and *Life Cycle Cost* calculations are also included in the assessment. Furthermore, social sustainability is equally prioritized with concepts of *User safety*, *Indoor Air Quality* and *Access for all*. This makes DGNB one of the most inclusive in regards to the different aspects of sustainability.

If properly documented and small adjustments in design and construction were made to account for the DGNB process, many buildings constructed may be near to qualifying for the lowest certificate level of silver under the Denmark modified DGNB scheme [3, 5].

According to Denmark Statistics [6], in 2015 construction was completed on over 5.5 million square meters of all building types in Denmark. Yet, according to the Green Building Council Denmark [7], there are only 30 DGNB certified buildings. Of those 30, 20 are office and administrative buildings. Thus, there appear to be many barriers to obtaining a certification. Several of them include:

- the cost of the certification and auditing;
- paperwork man-hours;
- additional cost required in the construction of the building;
- access to databases or the resources to generate the database;
- lack of basic understanding and the complexity of DGNB;
- and lack of general knowledge of the DGNB system [8].

Because of these barriers, it also appears that the largest firms have a significant advantage over smaller firms working on smaller projects. Though the largest projects may garner the most attention, the majority of the building stock in Denmark are homes and other smaller buildings, often completed by small or middle-sized companies, indicating a large potential market [9].

STAPLE, a companion to the DGNB International CORE 14 criteria manual, introduced in this paper is intended to help reduce some of the above-mentioned barriers.

2 UNDERSTANDING STAPLE

STAPLE (Sustainable Tool for Assessment, Planning, Learning, and Engaging) is an Excel-based interactive and iterative, education based platform intended to engage dialog among the Stakeholders & Decision-Makers and the assigned group Team Leaders, which are based on the five DGNB topics:

1. Economic Quality (ECO)
2. Environmental Quality (ENV)
3. Process Quality (PRO)
4. Sociocultural and Functional Quality (SOC)
5. Technical Quality (TEC)

STAPLE assesses the user's knowledge of sustainable buildings through several confidential questions, not revealed to others at any point when they first work with the tool. Based on the response and the level of knowledge, the tool provides additional educational material. At each stage of the tool, the user is provided plain-language easily digestible summaries of each Topic and Criterion. This is provided using multiple formats (PDFs, summary and bullet lists) and in different locations throughout the tool to aid in the learning process. Using different methods of input for differentiation, the users are prompted at two levels. The top-level input helps the user prioritize the DGNB five Topics. Each Topic is broken into Criterion, a tertiary level of DGNB. Each user is provided Topic-specific Criterion statements, a summary of the Criterion and asked to provide their understanding of the importance of that statement with regard to the other statements of that Topic. The assessment of the user input is provided graphically to encourage dialog. Visually seeing the difference of opinions among the users is intended to engage the users in discussion. The planning stage of the tool relies on the user's input of Confidence in Completion or their opinion of what level a specific activity might be completed for the project. This results again in a graphical output of the user's responses to engage the users in further dialog. Finally, with all input from all of the users and an estimation of DGNB indicator weighting, the planning of the project can be driven by an arranged list of indicators. If the list is followed, the highest concerns of the users will be met while possibly obtaining the most amount of DGNB points if properly documented.

The tool, tested at multiple stages of development and molded by many individuals both within and outside of the sustainable building industry, have shown, through their feedback that this is a desired product in the market. It achieves the primary goals of assessment of individual's current knowledge, educating through multiple stages and formats, the inspiring of conversation among team members through a graphical display of opinions, and helping the design team plan a project to possibly obtain the highest DGNB score.

2.1 Design process

According to various sources, the earlier sustainable strategies are considered and implemented into a project, the more successful the outcome [10]. This calls for a different approach as opposed to the conventional design process, Fig. 1, with a sketching phase carried out by the architect, a design phase in which an engineer gets involved, and an execution phase, in which contactors are appointed.

While there are several interpretations of an Integrated Design Process, as established by Camilla Brunsgaard, Mary-Ann Knudstrup, and Per Kvols Heiselberg in the paper *Experiences from the Design Processes of the First 'Comfort Houses' in Denmark*, The Integrated Design Process, Architecture & Design, Aalborg University (AOD IDP), which is an approach developed as a method for architecture students at Aalborg University, Civil Engineering in Architecture & Design, will be considered for STAPLE, as illustrated in Fig. 2 [11].

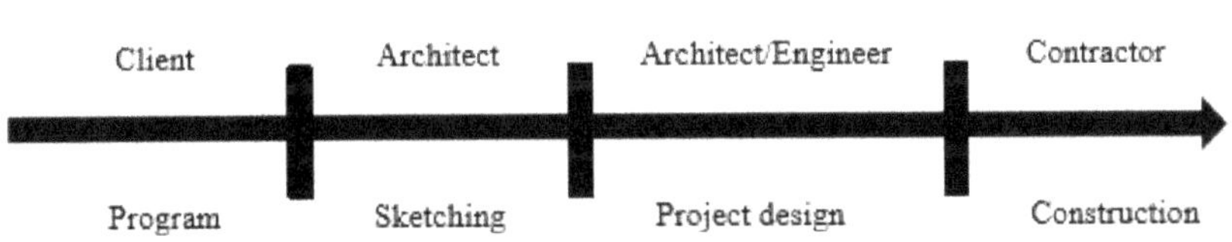

Figure 1: A traditional design process – a linear approach [11].

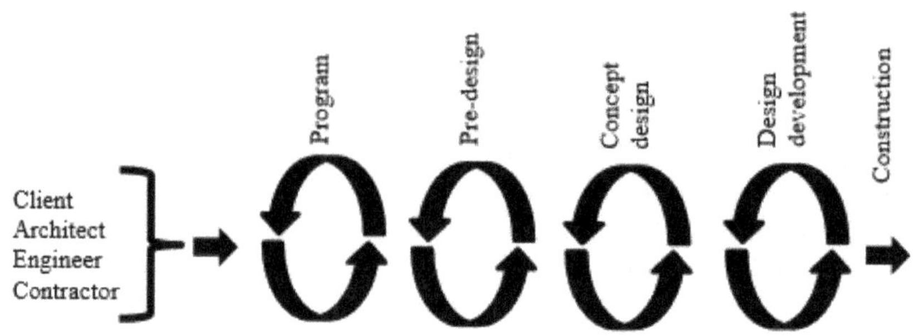

Figure 2: The Integrated design process – an iterative approach [11].

By repeating a cycle of design proposals, discussions, early changes, and subsequent new proposals, a design embodying the input and demands of all participants while fulfilling the overall project criteria, gradually emerges. STAPLE is suitable for an Integrated Design Process, as the live feedback generated by it allows the user to work in iterations and use the tool throughout the design process in various stages of the project.

2.2 STAPLE introduction

The education is intentionally provided in many ways. It is aimed at breaking the system down into plain language and small but manageable pieces.

The educational material is available throughout the tool as embedded PDFs that are always available to the user through the use of the STAPLE navigational bar that resides on the right side of the screen by default. Secondly, each user is then asked questions about their basic knowledge of sustainable buildings and the DGNB system. If the user is authentic with themselves and the tool, they may or may not be required to open some of the embedded educational material PDFs.

Third, the material is also broken down and provided in statements for each Topic. Each Criterion statement is presented as a plain language summary of comments that appear as a popover for a quick overview of what is entailed within each statement.

Once all input has been provided, the Stakeholders & Decision-Makers (SDM) are provided with results that compare their responses to the design Team Leaders (TL). SMD are intended to be building owners and their representatives, client consultants, and alike. TLs – often specialists within their respective fields, or professionals in charge of certain aspects of sustainability within a project team. Within STAPLE, there is one SDM and five TLs – each responsible for one of the five Topics within the DGNB. This is intended to create discussion between the groups. Providing independent input into the tool, an open discussion is encouraged to contemplate the differences of opinions. Having learned more about the system through the STAPLE process, the users may gain significant insight into their own understanding but also the understanding and knowledge of those working on the project team. This connection to the knowledge should help the communication of the team throughout the design process.

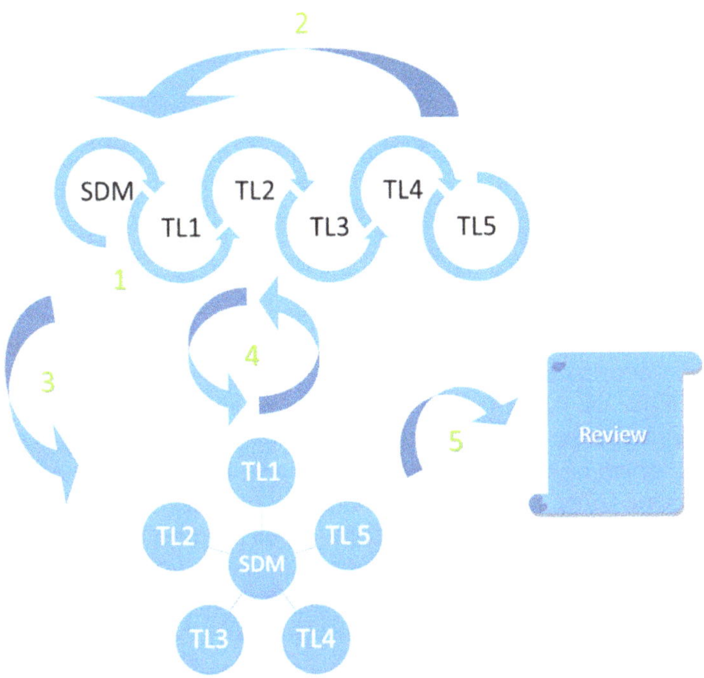

Figure 3: The intended flow of STAPLE sharing.

1. The Stakeholder and Decision Maker (SDM) starts up the tool, sets up PIN codes, determines the initial desired level of certification (if above Silver), and files out the respective sheets before saving the tool, exiting and passing it on to the Team Leaders (TLs).
2. Once the five TL input sections have been filled out, the tool would return to the SMD for review.
3. The SDM and all TLs discussing and reviewing the results.
4. The iterative process of several rounds of adjustments to the design until a satisfactory result has been achieved.
5. The final selection of STAPLE outcomes is reviewed by the project team members.

The final part or the advanced portion of the tool are prioritized lists of indicators, DGNB established design parameters needed to complete in order to achieve certification, based on the user input and an overview analysis of the possible DGNB certificate level. The analysis, based on either the Stakeholders & Decision-Makers or Team Leaders' Confidence in Completion (CiC) values and the DGNB estimated weights, a determination of possible certificate level is projected. The CiC value is the user's estimation of the level to which a certain design goal can be accomplished. Maximum amount of points possible in DGNB are awarded for a 100% completed criterion. The lists can be provided in several ways including the estimated indicator share of the total score if the DGNB weights are included. Otherwise, the user may create multiple lists depending on what weighting they would like to include in the system.

Through the focused interaction with STAPLE, commitment to reading the provided educational materials in all formats, and candid inputs, STAPLE has the ability to significantly reduce some of the barriers to entry into the DGNB Sustainable Building Certification system. The tool was tested many times by multiple people with a wide range of backgrounds. Their feedback is an indicator that this tool is helpful and can create an impact in the market if properly used.

STAPLE improves the user's general knowledge of the DGNB System and reduces the complexity to allow more individuals to be involved and learn about the process. STAPLE is just the first step in engaging a much wider audience to the understanding and implementation of the DGNB Building Certification System in many more projects.

2.3 DGNB system

Deutsche Gesellschaft für Nachhaltiges Bauen, DGNB, or the German Sustainable Building Council's system of sustainable building certification is based upon a hierarchal chain of information. At the top of the systems' chain are the five primary topics and one secondary topic. The five primary topics are the focus of this paper as the sixth topic, Site, does not influence the scoring and determination of certificate level. Under the Topic are the Criteria Groups; within each Criteria Group is a collection of Criterion and their associated Indicators. Each Indicator is a requirement that must be fulfilled to obtain the checklist points.

Although the DGNB assessment scheme has been investigated for the purpose of developing STAPLE and the math behind the evaluation matrix has been analyzed, the reasoning behind the DGNB weighting factors is unknown. The recreated matrix is, therefore, solely an estimation.

2.3.1 DGNB Performance indices

Performance Indices are the method of evaluation developed by DGNB to rate each building. To elaborate further, there are two parts of the performance index that must be satisfied to qualify for a certificate.

The first is the Minimum Performance Index (MPI). This is the requirement of each and all Topics. Accordingly, to be considered for a Silver certificate, all Topics must receive at least 35%. The second part of the qualification is the Total Performance Index (TPI), which is the overall requirement. At Silver, at least 50% of the total score must be awarded. If only one Topic does not qualify but the TPI is achieved, a certificate will not be awarded [12].

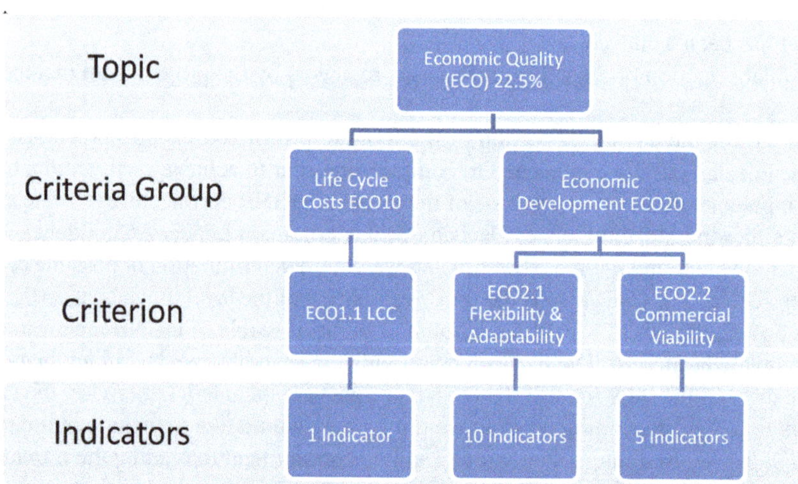

Figure 4: DGNB Topic hierarchy [4].

2.4 STAPLE user input

The five Topics within the system of sustainable building certification developed by DGNB have been assigned a weight, 22.5% to four Topics and 10% to the fifth Topic for a total of 100%. The reasoning behind this weighting is uncertain. This weight is their opinion of each Topic importance. In STAPLE, for input from the user, their opinion is desired and not influenced by the DGNB system. Therefore, before user input, the base value for all five Topics are equal or 20%.

For this project, there are five Topics. The user prefers each Topic but it is not clear to what percentage. For this reason, a pairwise comparison of the five Topics was made using the Analytic Hierarchy Process (AHP) as developed in the 1970's by Thomas Saaty [13]. Ten Pairwise comparisons of the five Topics are needed. Because of the limitations of AHP, it will only be used for the five Topic comparisons.

The Stakeholders & Decision Makers (SDM) and each of the Team Leaders (TL) are asked to compare the five Topics via the ten statements. The user is asked to slide the slider to a position that agrees with their opinion. When the slider is at the left, the left Topic is most heavily favored and the opposite when the slider is on the right. When the sheet is reset, the sliders are randomly placed on either the left or the right of the selection as to not influence the user. The right column provides feedback to the user regarding their selection. The explanation corresponds directly with the slider input. A consistency ratio is calculated concurrently with the user input.

In this application of the AHP, the deviation used in STAPLE is acceptable as it allows and encourages magnified differences between the Stakeholders & Decision-Makers and Team Leaders' inputs. If the Stakeholders & Decision-Makers place significant emphasis on the economics of the project yet all the Team Leaders are different, it will be visually obvious and may encourage discussion, one of the primary goals of the tool.

In addition, as the Team Leaders, the users are asked to provide a Statement Importance percentage for each of the provided statements within the five DGNB Topics that should amount to a total 100%. This is to indicate which statements the user believes are more important and is done through direct input.

The Input Comparisons look at the recorded Statement Importance (SI) from each Team Leader (TL) and the Stakeholders & Decision Makers (SDM). Both the Stakeholders & Decision-Makers as well as the Team Leaders were required to provide their opinion on the same statements. Each Team Leader is only required to answer Topic-specific statements while the SDM is required to answer all of the statements.

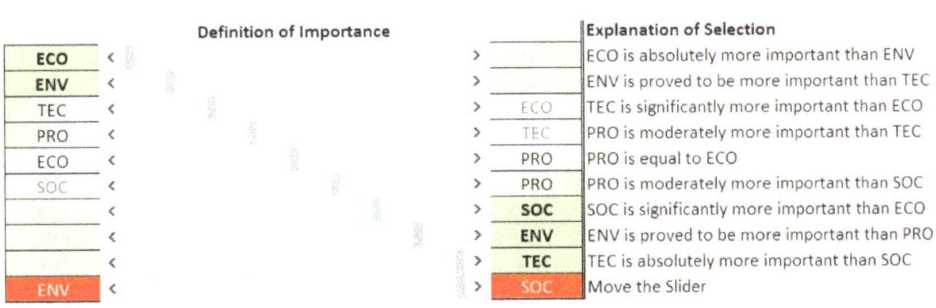

Figure 5: Pairwise user input in STAPLE.

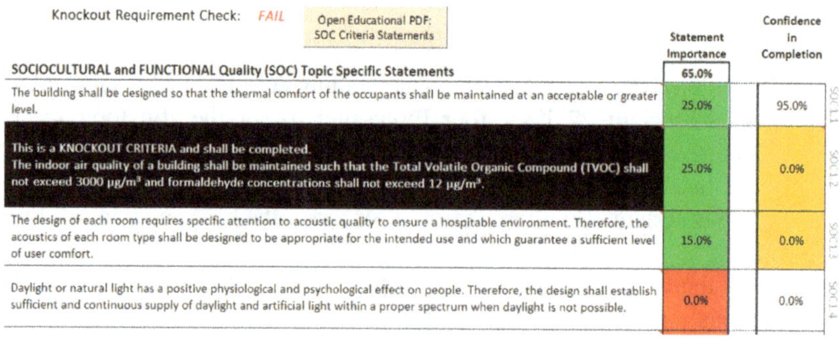

Figure 6: STAPLE Team Leader Input with Statement Importance (SI) and Confidence in Completion (CiC).

Then the Stakeholders & Decision-Maker was required to input a Confidence in Completion (CiC) for each Topic while the Team Leaders were required to provide a detailed input for each Criterion. Some CiC values have been set minimum limits, according to the DGNB requirements.

2.5 STAPLE output

DGNB Indicator weighting is required to estimate a score and the associated list of indicators. The Stakeholders & Decision-Makers provide a ranking of the five Topics with a pairwise AHP comparison. This is the first sorting of all the Indicators. Two lists are provided. The first list is based on the Team Leaders' input while the second is based on the Stakeholders & Decision-Makers. With the DGNB Indicator weight or share of the total score, the Stakeholders & Decision-Makers/Team Leaders' CiC is applied. This is then considered the *adjusted indicator weight* and will be used to fulfill the requirements of the Minimum Performance Index (MPI) and Total Performance Index (TPI) as described earlier.

The sorting begins with the highest Stakeholders & Decision-Maker's ranked Topic. The highest ranked is filled with adjusted indicators until the minimum level for the requested certificate level is achieved. Once filled, the sorting continues to the second-ranked Topic to complete. This continues until all five Topics are completed. The higher the desired certification class, the more adjusted indicators are required. Once each of the Topics has enough adjusted indicators to complete the MPI, the Total Performance Index is investigated.

The sum of the indicators used to complete the Topic is collected and compared to the desired certificate level's TPI. If the TPI is not met, additional indicators are needed. The sort selects adjusted indicators with the highest total share from the list of indicators not already used to complete the Topics. Adjusted indicators are added until the TPI is obtained. If the TPI is not possible because of low CiC, all indicators will be shown.

For each Stakeholders & Decision-Makers and Team Leaders' List of Indicators, an output similar to that in Fig. 6 is provided. At quick review to the right side, the user can assess if the desired level of certificate is possible. The Estimated TPI is determined by summing all the adjusted Indicator's share of the total. This is then presented and compared to the desired certificate. To the right of the TPI estimate is the certificate, which that portion of the score would qualify.

	Team Leader *Estimated* Total Performance Index		
Estimated:	75%	Gold	Performance Ratio, Est/Min:
Minimum:	50%		= 75% / 50% = 1.50

	Team Leader *Estimated* Topic Minimum Performance Index				
	TEC	ECO	PRO	SOC	ENV
Estimated:	98%	75%	98%	66%	73%
Minimum:	35%				
Estimated Certificate:	Platinum	Platinum	Platinum	Platinum	Platinum

Desired Certificate: Silver

Possible Certificate: Gold

Figure 7: STAPLE Example Estimated Performance Index (Silver).

	Team Leader *Estimated* Total Performance Index		
Estimated:	75%	Gold is possible. Does Not	Performance Ratio, Est/Min:
Minimum:	80%	Meet Desired Certificate Level	= 75% / 80% = 0.94

	Team Leader *Estimated* Topic Minimum Performance Index				
	TEC	ECO	PRO	SOC	ENV
Estimated:	98%	75%	98%	66%	73%
Minimum:	65%				
Estimated Certificate:	Platinum	Platinum	Platinum	Platinum	Platinum

Desired Certificate: Platinum

Possible Certificate: Gold

Figure 8: STAPLE Example Performance Index (Platinum).

To the right of the estimated possible certificate for the TPI is a performance ratio. This helps determine how well the TPI is performing compared to the desired level. This is helpful in determining how much effort might be needed for the next level certificate.

Below the TPI information is the MPI for each Topic. Similar to that of the TPI, each Topic's adjusted Indicators are summed based on the share of the Topic weight. This is concisely presented. The MPI for the desired certificate is presented. The final options shown are the possible certificate level for each Topic.

For this example, the Total score is the limiting factor. If the desired level is changed to a level that is not possible, the table will update. Figure 7 shows that the table provides a visual clue that this certificate level does not qualify by marking which element is not sufficient.

2.6 Indicator Relative Power Ranking

STAPLE has five modifiers from the users:

1. Stakeholders & Decision-Makers Topic Ranking Percentage
2. Stakeholders & Decision-Makers Criterion Group Confidence in Completion
3. Stakeholders & Decision-Makers Criteria Statement Importance
4. Team Leaders Criterion Confidence in Completion
5. Team Leaders Criteria Statement Importance

The analysis of the DGNB System discovered the Indicator weight as a share of the total score.

These six mutually exclusive weight and modifiers can be combined to modify the rank of each criterion. If the DGNB weight is included, the detail increases to the indicator

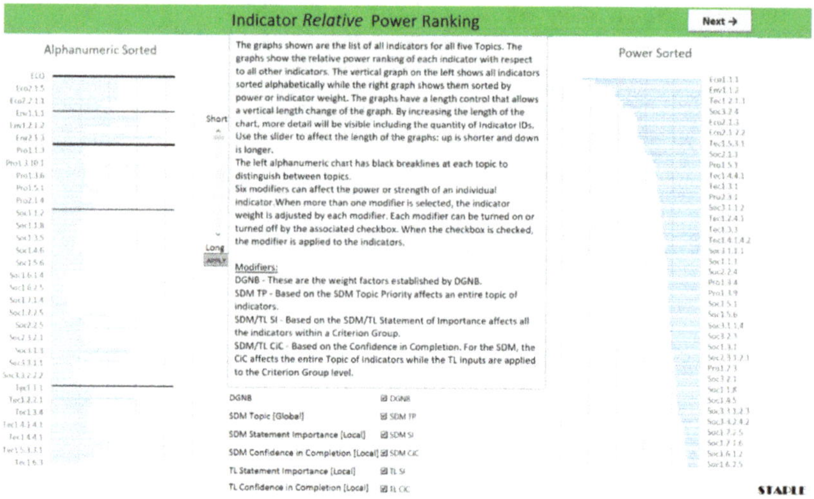

Figure 9: Indicator Relative Power Ranking, STAPLE results, screenshot.

level. The Indicator Relative Power Ranking allows the user to view the list of Indicators sorted in two ways. The first list is alphanumerically sorted with ECO1.1 shown first and indicators in TEC shown last. The second list is sorted by relative power. Below, an illustration of this is provided. The user is also given explanation and guidance in order to interpret the results.

3 CONCLUSION

An excel-based tool, named STAPLE (Sustainable Tool for Assessment, Planning, Learning, and Engaging) was initially developed and is intended to assist various project team members in decision making and work-sharing process, to facilitate the Integrated Design Process and to encourage and promote discussions throughout the project phases. However, through the research and development of the tool, the importance of education became abundantly clear and is highly integrated throughout.

Several issues, characteristic to building certification and sustainability in the building industry in general, illuminated key components, which would become part of STAPLE in order to offer a guiding and supporting tool. One of the largest barriers to the use of DGNB is the complexity and understanding of the base material. STAPLE solves this by providing the user a breakdown of the primary elements of DGNB. The purpose of STAPLE is to facilitate the design process, encourage discussions, educate the user on various aspects of sustainability, and visually present the user input (opinion) from various project team members, which further promotes dialogue between the parties. A great deal of effort and time has been dedicated to the educational aspect of STAPLE. Explanations, directions, and guidelines have been added to each of the software sheets to help the user navigate through the tool. Many educational documents have been embedded into STAPLE as print-friendly PDF files. These are compactly formatted, abridged bite-sized summaries of various aspects of sustainability, breaking down information regarding DGNB in easy-to-read, easy-to-digest portions. Some documents overlap information from other documents to help the user see the 'red thread' of connectivity between concepts. Throughout the investigation, the information gathered has been presented in a simplified manner, providing the basics, useful links, and sources for further information, if desired. It is also worth noting, that the statement 'hover-over' or 'pop-up' comments, written for the laymen perspective, embedded into STAPLE, further supports the user understanding and is intended to provide the user with an overview of many different aspects of sustainability and the DGNB evaluation matrix through several 'digestible' pieces of information at a time. Besides promoting dialogue, one of the primary goals has been to change the perception of sustainability being an overwhelming and utterly complex topic by presenting the information in a simple and straightforward way piece by piece. Additionally, there are several other documents found in STAPLE, such as the *Sustainability contract* and the *Criteria statement supplements*. Based on the user feedback, the general response regarding the provided educational materials and user guides was largely positive with the majority of testers finding the information helpful, easy to read, and interesting.

As presented, STAPLE shall be considered a working prototype. It needs to be professionally upgraded to be more useful in the market. These upgrades would be to increase functionality to include beginner, intermediate, and advanced modes, improve the user's interface, include reporting capabilities, end reduce file size through cloud storage of PDFs. It is clear there is a market for a smaller, gentler version of DGNB for the much smaller projects. STAPLE, together with organizations like the Green Building Council Denmark, can accomplish a greater number of investors, buildings, designers, and owners being engaged, educated, and converted into using DGNB while developing high-performance quality buildings.

ACKNOWLEDGEMENTS

This paper became a reality with the kind help and support of many individuals. We would like to take the opportunity and extend our sincere thanks to all of them.

Foremost, we are grateful to the teaching staff at Aalborg University, whose expertise, understanding, guidance, and support made it possible for us to work on a topic that was of great interest to us.

We are highly indebted to the skilled professionals, who volunteered to test STAPLE throughout its various stages of development. Without their responsive and helpful feedback, STAPLE would have never advanced to where it is today.

We would like to express our gratitude to Hicham Johra, Aalborg University, whose feedback and guidance through the project left an invaluable impact on STAPLE and helped improve it beyond our own efforts.

We would also like to thank Gitte Gylling Hammershøj Olesen, Rambøll Aarhus, for her involvement, professional interest, and helpful feedback from the very beginning of this thesis.

We are also thankful to the adept and kind individuals, who, through informal conversations, support, and advice, helped shape this report. There are too many to name.

Our thanks and appreciations go out to everyone, who have willingly helped us with their knowledge.

REFERENCES

[1] Learn About Sustainability; United States Environmental Protection Agency Online. Available at: https://www.epa.gov/sustainability/learn-about-sustainability#what. (Assessed 2 January 2017).

[2] Why the building sector?; Architecture 2030, Santa FE Online. Available at: http://architecture2030.org/buildings_problem_why/#lightbox[group-158]/0/. (Assessed 2 January 2017).
[3] Danish Building Regulations 2015; The Danish Transport and Construction Agency Online. Available at: http://bygningsreglementet.dk/english/0/40. (Assessed 9 September 2016).
[4] The DGNB sustainability concept. The new quality of building; DGNB GmbH, Stuttgart, Deutschland Online. Available at: http://www.dgnb-system.de/en/system/dgnb-sustainability_concept/. (Assessed 16 November 2016).
[5] An introduction to DGNB; Green Building Council Denmark (DK-GBC), Copenhagen, Denmark. Available at: http://dk-gbc.dk/media/2292/dgnb_dk-gbc_oct_2012.pdf. (Assessed 22 November 2016).
[6] Nielsen E., Statistics Denmark. Available at: http://www.statbank.dk/BYGV01. (Assessed 09 January 2017).
[7] Certified projects; Green Building Council Denmark (DK-GBC), Copenhagen, Denmark. Available at: http://www.dk-gbc.dk/dgnb/certificerede-projekter/. (Assessed 09 January 2017).
[8] Hankinson M. & Breytenbach A., *Barriers that impact on the implementation of sustainable design*, Helsinki, Finland, 2012. Available at:http://www.unece.lsu.edu/greenbuilding/documents/2013Mar/gb13_04.pdf. (Assessed 25 November 2016).
[9] Byggeri og anlæg; Statistics Denmark, Copenhagen, Denmark. Available at: http://www.dst.dk/da/Statistik/emner/erhvervslivets-sektorer/byggeri-og-anlaeg. (Assessed 12 December 2016).
[10] Benefits for all; DGNB GmbH. Stuttgart, Deutschland. Available at: http://www.dgnb-system.de/en/certification/benefits/. (Accessed 16 November 2016).
[11] Brunsgaard, C., Knudstrup, M.A. & Heiselberg, P.K., *Experiences from the Design Processes of the First "Comfort Houses" in Denmark*: Aalborg, 2009.
[12] Evaluation and awards; DGNB GmbH. Stuttgart, Deutschland. Available at: http://www.dgnb-system.de/en/system/evaluation_and_awards/. (Assessed 12 December 2016).
[13] Saaty, T., *How to make a decision: The Analytic Hierarchy Process* (Report No.: 0377-221), 1990.

APPLICATIONS OF LOW ENTHALPY GEOTHERMAL ENERGY: THE CASE OF THE FACULTY OF PHYSICAL AND MATHEMATICAL SCIENCES AT THE UNIVERSITY OF CHILE

CLAUDIA MAC-LEAN, REYNALDO CABEZAS, MAURICIO MUÑOZ & LUIS VARGAS
Faculty of Physical and Mathematical Sciences, University of Chile, Chile.

ABSTRACT

In recent decades, the industry has observed a significant shift towards the use of renewable energy, such as solar, wind and geothermal. The Chilean scenario has not been an exception, and much progress has been made in sustainable energy prospection and implementation, especially in the electricity sector, where solar and wind power amount 2300 MW, and since April 2017, the first geothermal power plant (48 MW) has come into operation. In the area of low enthalpy geothermal energy, the use is around 19 MW.

The Faculty of Physical and Mathematical Sciences at the University of Chile has been contributing to this transformational process, with its Sustainable Campus initiative. The first step of this initiative is the introduction of renewable energy on site, which has been achieved through the installation of a solar photovoltaic plant of 15 kW. Along this line, the design and implementation of a geothermal air conditioning system (HAVC) is underway, which will serve the classrooms and offices in the traditional engineering building of the campus. The technology to be used in this project is the Ground Heat Pump (GHP).

The present paper includes an introduction of the applications of low enthalpy geothermal energy in Chile, a description of the Office of Engineering for Sustainable Development at the Faculty of Physical and Mathematical Sciences, and the design of a geothermal HAVC system in the university campus, considering economic, environmental, technical and social aspects. Besides the operation of the GHP, the system will be used for teaching purposes to incorporate sustainable development in the curriculum of the university. The expected savings of the geothermal system versus an aerothermal design are 41,070 kWh annually, considering both cooling and heating.

Keywords: Geothermal energy, renewable energy, sustainable university campus.

1 INTRODUCTION

Geothermal Heat Pumps (GHP) systems are one of the fastest growing renewable technologies in the world, with most of this growth happening in China, USA and Europe. The number of GHP increased ~10% annually between 2010 and 2015. By the end of 2015, the global installed capacity was estimated at almost 50,258 GWth with an annual energy use of 326,848 TJ/year. This represents 70.9% of installed capacity and 55.15% of the use of direct uses of geothermal energy [1].

Santiago is the Chilean capital city, where more than 6.5 million inhabitants are living. The city has pollution problems, which are increased in winter time by the use of conventional heating systems that produce emissions and particulate matter. On the other hand, in summer time, temperature reaches up to 30–35°C and therefore, cooling systems are required. Typical Santiago building cooling systems have low efficiency and high costs of operation and maintenance.

In this work, the use of GHP is proposed for heating and cooling a building in the Faculty of Physical and Mathematical Sciences (FCFM for its acronym in Spanish) at the University of Chile. The GHP is operated with electricity and therefore there is no direct emissions of

DOI: 10.2495/EQ-V3-N1-69-78

particulate matter or greenhouse gases during the system functioning. The GHP concentrates and transports heat from the ground to the building, and vice versa, for heating and cooling purposes. This technology represents savings of 25–50% over air source heat pumps [2].

Due to the excellent aquifer features at the FCFM, an open loop system is proposed to feed the Ground Water Heat Pump (GWHP). Operation of a GWHP consists of extracting groundwater to exchange heat and, subsequently, the water is re-injected into the aquifer [3].

In Chile, geothermal heat pumps are scarce because of the lack of knowledge of the benefits of this type of heating and cooling system, and high initial investment due to drilling or excavation costs. However, as background for the proposed system at the FCFM, similar case studies in the international and national context are summarized in the following table.

To carry out the analysis of technical and economic sustainability, the operation of this heating and cooling system is modelled for a section of a building at the FCFM. The performance of this system is compared against a standard aerothermal HAVC system in order to show the savings of the geothermal solution. This case is expected to promote the use of this technology for similar buildings on campus.

2 ENERGY NEEDS AND PROJECT METHODOLOGY

2.1 Energy needs at university campus

Founded in 1842, the University of Chile is the main and oldest institution of higher education owned by the State in Chile, with a national and public character. It is a research-oriented institution, and it is organised in 14 faculties.

The FCFM, in which our work is focused, hosts the School of Engineering. It has 425 professors, 1100 graduate students and 4860 undergraduate students.

Table 1: International and National Similar Projects.

Place	Brief description	Capacity
Bonn, Germany	20 households	450 kW of heating capacity from underground water
Frankfurt, Germany	Building with offices	1 MW heating, 4 wells
Bonn, Germany	United Nations Campus	4.5 MW, 3 wells
Baku, Azerbaiyán	Diplomatic Academy of Azerbaiyán, heating and cooling	2000 kW heat, 1200 kW cold, 290×140 m of geothermal probes
Dortmund, Germany	Buildings with offices, heating and cooling	800 kW heat, 600 kW cold
Cologne, Germany	'KidZ' service centre, heating and cooling	150 kW, underground water used
Linz, Austria	Sports centre, heating and cooling	170 kW, 26×100 m of geothermal probes
Quelen Hotel, Chile	Heating and cooling system for the building	40 kW with a Coefficient of Performance COP of 5
Hermanita de los pobres residencie, Chile	Heating of a residence for adults	100 kW with a Coefficient of Performance COP of 4

In terms of energy consumption of the FCFM, it is relevant to present certain preliminary information, in order to start dimensioning the challenge and the project itself. In 2015, the campus spent 7.6 GWh of electricity which represents approximately USD$ 1 million. Figure 1 shows the electric grid of the campus.

In terms of consumption, the main areas are heating, ventilation and air conditioning (HVAC), informatics, lighting, labs, and others. The share of this demand is shown in Fig. 2.

Specifically, the informatics consumption is composed of 50% by servers and 50% by desktop computers. In terms of lightening, 80% of it is by fluorescent tubes on campus. This consumption contemplates seven buildings, one of which is the 'Edificio Escuela', where the geothermal system for heating and cooling is proposed.

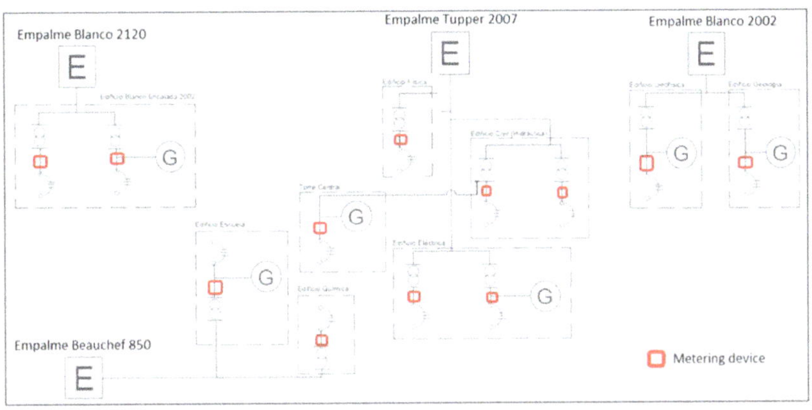

Figure 1: Electric grid and connecting points at the FCFM.

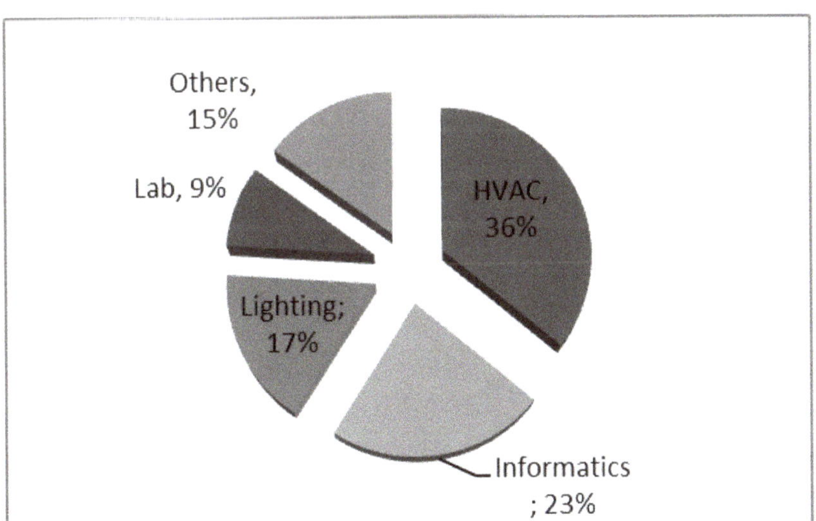

Figure 2: Electric consumption of the FCFM.

2.2 Methodology

Regarding the typical analysis for heating and cooling systems with geothermal heat pump (4), the methodology to undertake this project has the following stages:

A. Estimation of the heating and cooling loads, taking into account local exterior temperature, comfort temperature of the classrooms and offices in the building, design features of the building, air renovations and solar gains.
B. Estimation of thermal energy usage based on heating and cooling loads and hours used throughout a year.
C. Determine wells requirements based on groundwater flow to guarantee nominal function conditions of the water to the heat pump.

In the next section, the methodology steps for the project are presented.

3 PROJECT DESCRIPTION AND RESULTS

3.1 Estimation of heating and cooling loads

The main heating and cooling loads of this project maybe classified in six categories:

3.1.1 Outside temperature and comfort temperature inside the building

A key element to calculate the thermal demands, are the temperatures in the zone where the system is to be installed. Typically, minimum and maximum temperatures are considered. In the present study, data is used from 'La Platina' meteorological station, taken with 1 h of difference between years 2012 and 2015. Data shows that the minimum temperature in winter time is ~1°C and maximum temperature in summer time is ~32°C. In addition, daily thermal amplitude is ~10°C in winter time and ~15°C in summer time (Fig. 3).

The maximum temperature in the day is a round 14:00 and 16:00 h, when the building is used at full capacity. On the other hand, the minimum between 04:00 and 06:00, when only

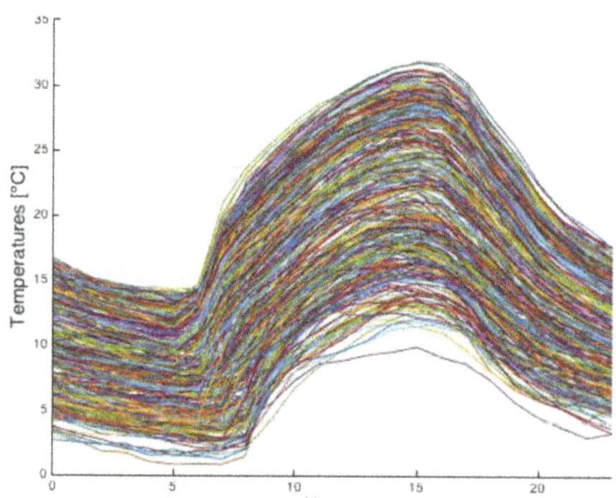

Figure 3: Daily temperatures in a typical year in Santiago, Chile (average of the years 2012–2015).

the security staff is at the campus. For heating and cooling purposes, the inside temperature for the building is defined as 24°C and 20°C, respectively.

3.1.2 Identification of the places (classrooms, offices, halls) to be intervened and their global heat transfer coefficient

The current heating and cooling systems in the building to be intervened 'Edificio Escuela' is an air source heat pump used only for cooling purposes (model 20 GTN 080), which dates from 1996 and some of them correspond to isolated heating electric devices. The heating/cooling devices are 12 air handlers with an electric heating source and 5 fancoils. Such equipment provides services for several classrooms, offices, and an auditorium. All are located in the south sector of the 'Edificio Escuela'.

The present projects aims at modernising these installations, and additionally integrate new areas, in order to provide heating and cooling for all the classrooms and offices located in the south part of the 'Edificio Escuela'. The main clients include 10 classrooms and several offices.

The size and dimensions of the different zones to be heated and cooled by the geothermal heat pump (comprising classrooms and offices) are detailed in Table 2. The heat transfer

Table 2: Heat transfer coefficient for offices and classrooms to be intervened, according to their size, materials and heat transfer coefficient (U), considering error for measuring

Classroom/office	**Window [m²]**	**Wall [m²]**	**UA window [W/K]**	**UA wall [W/K]**	**UA total [W/K]**
S14	24.2±0.3	48.5±0.3	132.8±7.8	121.1±12.9	254.0±20.7
S15	28.5±0.3	94.8±0.3	156.7±8.8	236.9±24.5	393.6±33.4
S16	8.1±0.3	14.4±0.3	44.3±3.7	35.9±4.4	80.2±8.2
S17	12.8±0.3	33.1±0.3	70.5±4.9	82.7±9.1	153.2±14.0
S18	8.1±0.3	14.4±0.3	44.3±3.7	35.9±4.4	80.2±8.2
S19	28.5±0.3	94.4±0.3	156.7±8.8	235.9±24.4	392.6±33.3
Treasury	16.1±0.3	33.6±0.3	88.7±5.8	84.0±9.2	172.7±15.0
S24	20.7±0.3	51.9±0.3	113.9±6.9	129.8±13.8	243.6±20.7
S25	24.4±0.3	98.8±0.3	134.3±7.8	247.1±25.5	381.4±33.4
S26	11.0±0.3	38.7±0.3	60.4±4.5	96.7±10.5	157.1±15.0
S27	7.0±0.3	18.2±0.3	38.4±3.5	45.6±5.4	84.0±8.9
S28	24.4±0.3	98.4±0.3	134.3±7.8	246.1±25.4	380.4±33.3
S29	10.5±0.3	47.0±0.3	57.8±4.4	117.6±12.6	175.3±16.9
Summer School	13.8±0.3	35.3±0.3	75.9±5.2	88.4±9.7	164.3±14.8
S30	0.0±0.0	48.3±0.0	0.0±0.0	120.7±12.9	120.7±12.9
Off.1 3rd Floor	10.1±0.3	38.2±0.3	55.7±4.3	95.6±10.4	151.3±14.6
Off.2 3rd Floor	3.9±0.3	5.1±0.3	21.5±2.7	12.8±2.1	34.2±4.8
Off.3 3rd Floor	7.8±0.3	11.7±0.3	42.9±3.7	29.1±3.7	72.0±7.4
Off.4 3rd Floor	5.1±0.3	5.4±0.3	28.2±3.0	13.6±2.2	41.8±5.2
Off.5 3rd Floor	10.2±0.3	38.4±0.3	55.9±4.3	95.9±10.4	151.8±14.7

coefficient (U) for window and walls are equal to 5.5±0.25 [W/m²K] and 2.5±0.25 [W/m²K], respectively.

The global heat transfer coefficient for all the classrooms and offices yields a value of 3,684±335 [kW/K].

3.1.3 Transmission loads

The transmission loads contain all the heat losses or gains associated with the conduction and convection through the building surrounding. Transmission loads are calculated with the following equation:

$$Q_t = \sum_j U \cdot \Delta T \tag{1}$$

where U is the global heat transference coefficient [kW/K], ΔT is temperature difference between the outside and the comfort temperature for heating and cooling purposes. Transmission loads for summer and winter time are depicted in Fig. 4.

3.1.4 Air renovations loads

Thermal loads are those related to air renovations and subsequently are gain or losses if outside temperature is higher or lower than comfort temperature. It is determined by the following equation:

$$Q_v = \rho \cdot C_p \cdot \dot{V} \cdot \Delta T \tag{2}$$

where ρ is air density (1,2 [kg/m³]), C_p is the air specific heat (1.005 [J/kg·°C]), and $\dot{V}$ is air volumetric flow (5 [L/s] per person). Figure 5 shows the thermal ventilation loads in a typical summer day (blue) and winter day (red). The negative values mean heat losses and the positive values indicate heat gains. It is considered that the ventilation systems work from 08:00 to 19:00 h and are used at 70% capacity.

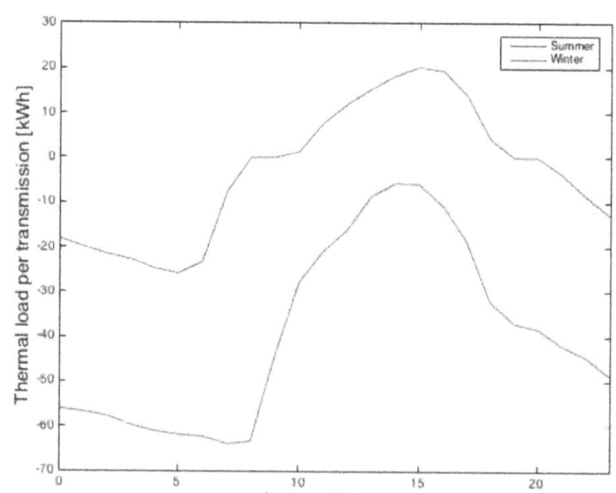

Figure 4: thermal transmission loads in a typical summer day (blue) and winter (red) (average of the years 2012–2015).

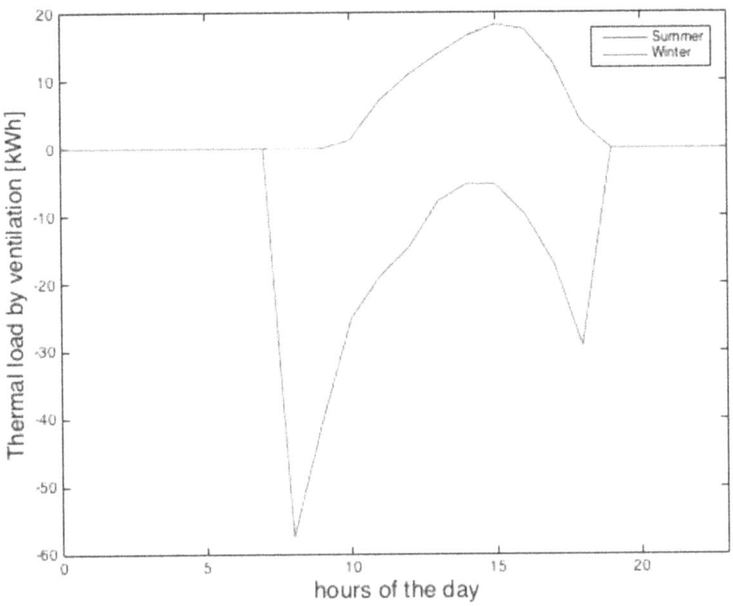

Figure 5: Air renovation loads in a typical summer day (blue) and winter (red), considering that the students are in university between 08:00 and 19:00 h (average of the years 2012–2015).

3.1.5 Internal gains for people and equipment

In the classrooms was considered: 1 projector (200 [W]), 1 computer (300 [W]), and all of lightening in each place. For the offices was consider 10 computers of 300 [W] and 96 lightening of 60 [W]. The working hours are from 08:00 to 19:00 h. The results are an amount of 15.2 [kWh] energy per hour of internal gains.

In terms of internal gains from people, an average person releases an estimate of 100 [W]. The maximum capacity between 08:00 and 19:00 is approximately 700 people. We consider in the range of 70% of the maximum capacity, which corresponds to 490 people. Thus, the heat gain is 49 [kWh]. Year round, the total internal gain is 64.2 kWh/h.

3.1.6 Solar gains

Solar gains are related to heating by sun light. For the present estimations since these gains were due to location of the south part of the 'Edificio Escuela', these solar gains are difficult to calculate.

3.2 Annual energy consumption

The resulting heating and cooling demand in MWh are summarized in Table 3. In addition, there is an estimation for the electric energy consumption taking into account heating and cooling COP (coefficient of performance) of geothermal and pumps and air source heat pumps. The operation of an Geothermal Heat Pump (GHP) is characterised by the coefficient of performance defined as the ratio between useful thermal energy E_t[Watt] and electrical energy consumption E_el [Watt]:

COP=E_t/E_el".

Table 3: Heating and cooling consumption relative to electrical consumption for a throughout a year (average of the years 2012–2015).

	Heating consumption	**Electrical consumption (GHP)***	**Electrical consumption (ASHP)****	**Cooling consumption**	**Electrical consumption (GHP)***	**Electrical consumption (ASHP)****
Month	**MWh$_{thermal}$**	**MWh$_{electrical}$**	**MWh$_{electrical}$**	**MWh$_{thermal}$**	**MWh$_{electrical}$**	**MWh$_{electrical}$**
Jan	5.80±0.52	1.32±0.12	1.61±0.14	27.89±2.51	4.98±0.45	9.96±0.90
Feb	5.98±0.54	1.36±0.12	1.66±0.15	24.98±2.25	4.46±0.40	8.92±0.80
Mar	9.03±0.81	2.05±0.18	2.51±0.23	24.87±2.24	4.44±0.40	8.88±0.80
Apr	14.37±1.29	3.27±0.29	3.99±0.36	15.87±1.43	2.83±0.25	5.67±0.51
May	19.29±1.74	4.38±0.39	5.36±0.48	9.00±0.81	1.61±0.14	3.21±0.29
Jun	23.57±2.12	5.36±0.48	6.55±0.59	4.15±0.37	0.74±0.07	1.48±0.13
Jul	26.04±2.34	5.92±0.53	7.23±0.65	2.96±0.27	0.53±0.05	1.06±0.10
Ago	22.70±2.04	5.16±0.46	6.31±0.57	4.50±0.41	0.80±0.07	1.61±0.14
Sep	17.29±1.56	3.93±0.35	4.80±0.43	7.03±0.63	1.26±0.11	2.51±0.23
Oct	14.45±1.30	3.28±0.30	4.01±0.36	13.68±1.23	2.44±0.22	4.89±0.44
Nov	10.88±0.98	2.47±0.22	3.02±0.27	19.91±1.79	3.56±0.32	7.11±0.64
Dec	8.21±0.74	1.87±0.17	2.28±0.21	24.93±2.24	4.45±0.40	8.90±0.80
Total	**177.61**±15.98	**40.37**±3.63	**49.34**±4.44	**179.77**±16.18	**32.10**±2.89	**64.20**±5.78

*GHP: Geothermal Heat Pump; **ASHP: Air Source Heat Pump

On one hand, typical values for heating and cooling COP of geothermal heat pump are 4.4 and 5.6, respectively [5]. On the other hand, typical values of heating and cooling COP of air source heat pump are 3.6 and 2.8, respectively [6]. Figure 6 shows the expected monthly cooling and heating electrical energy consumption.

3.3 Determine wells requirements

Given that the heating and cooling COP are 4.4 and 5.6, the heating and cooling loads are 130 and 136 kW, respectively. The maximum heat power extracted from the ground by the groundwater is ~100 kW (power from the groundwater minus power from the compressor at the heat pump). On the other hand, the maximum heat dissipated into the ground by the groundwater is ~160 kW (power from the building plus power from the compressor at the heat pump). The temperature difference between pumping and reinjection groundwater that thc hcat pump uses for functioning is 5°C. Finally, as the water heat capacity is ~4.18 kW/kg°C, the maximum pumping/rejecting rate is 5 lt/s and 8 lt/s for heating and cooling, respectively.

Statistics from the Chilean Directorate General of Water (Dirección General de Aguas, DGA) indicates that wells located close to the Faculty pump more than 10 lt/s. Therefore, one pumping well and one rejection well is enough for the open loop of the GHP.

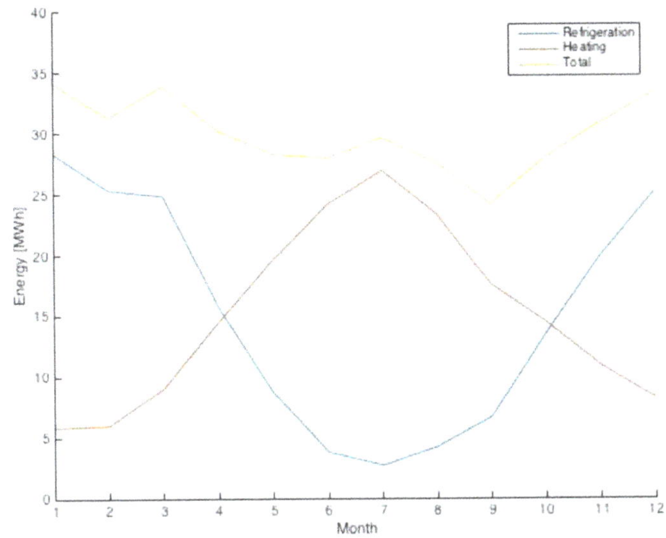

Figure 6: monthly cooling (blue) and heating (red) electrical energy consumption for air-condition with Geothermal Heat Pump (average of the years 2012–2015).

4 COMPARISON OF AEROTHERMAL AND GEOTHERMAL SYSTEMS

In this section, a comparison of aerothermal and geothermal systems is performed. In addition, a preliminary economic evaluation, taking into account differences between the initial investment of geothermal heat pump and air source heat pump. Also, an analysis of the savings in operation cost is presented.

4.1 Comparison of aerothermal and geothermal systems

The main difference between geothermal heat pump and air source heat pump is the COP due to efficiency in evaporator of the heat pump. Typical values for heating and cooling COP of geothermal heat pump are 4.4 and 5.6, respectively. On the other hand, typical values of heating and cooling COP for air source heat pump are 3.6 and 2.8, respectively (6). With the above data, the main saving operation cost is due to cooling with GHP.

4.2 Economic project evaluation

In the present chapter, the additional initial investment for the geothermal project will be presented, along with the expected savings in energy consumption.

The initial upfront cost includes the heat pump and the well implementation, for pumping and reinjecting water. On the one hand, the pump has to be included both in the aerothermal and geothermal project designs, so it does not imply additional spending. On the other hand, the specific additional cost will be in the wells. In Santiago, this costed US$ 85,000.

The estimated annual energy consumption of an aerothermal system is 113.54 MWh, whereas the geothermal option represents 72.47 MWh. Considering the cost of energy with the specific tariff of the Faculty of 77 $/kWh (0.11 US$/kWh), the expected annual saving should be US$ 4,520. Thus, the payback of the geothermal project is estimated to be 18 years.

5 CONCLUSIONS

This paper shows that higher savings are found when cooling rather than heating capabilities are used within a geothermal GHP system. The savings should increase when solar gains are included in the analyses.

Finally, the implementation of two wells might be enough to expand the infrastructure being heated and cooled at the FCFM, which might ameliorate the economic project evaluation.

ACKNOWLEDGEMENTS

This research was partially funded by the Andean Geothermal Center of Excellence (CEGA) and the Complex Engineering Systems Institute, ISCI (Project CONICYT: FB0816).

REFERENCES

[1] Lund. J.W. & Boyd. T.L., Direct utilization of geothermal energy 2015 worldwide review. *Geothermics*, **60**, pp. 66–93, 2016. Available at: http://www.sciencedirect.com/science/article/pii/S037565051500156X. https://doi.org/10.1016/j.geothermics.2015.11.004

[2] US. Department of Energy. 2011. Guide to Geothermal Heat Pumps.

[3] Self, S.J., Reddy, B.V. & Rosen, M.A., Geothermal heat pump systems: Status review and comparison with other heating options. *Applied Energy*, **101**, pp. 341–348, 2013. Available at: http://www.sciencedirect.com/science/article/pii/S0306261912000542. https://doi.org/10.1016/j.apenergy.2012.01.048

[4] ASHRAE. 2009. *Guide to fundamentals*. Atlanta. USA: American Society for Heating Refrigerating and Air Conditioning Engineers.

[5] Dimplex-SI130TUR+, 2017. Catálogo bomba de calor geotérmica Dimplex SI 130TUR+, p. 4.

[6] Dimplex-LA60TUR+, 2017. Catálogo bomba de calor aerotermia Dimplex-LA-60TUR+, p. 4.

Author index

www.ingramcontent.com/pod-product-compliance
Lightning Source LLC
LaVergne TN
LVHW062240200926
848484LV00006B/1581

* 9 7 8 1 7 8 4 6 6 3 1 9 3 *